Unveiling Cells' Inner Cleanup: Autophagy

Nama

TABLE OF CONTENTS

CHAPTER 1. AUTOPHAGY: A PROCESS ESSENTIAL FOR CELLULAR HOMEOSTASIS AND SURVIVAL

Macroautophagy (here referred to as "autophagy") is a highly conserved cellular process used to degrade misfolded proteins and damaged organelles. It also serves to recycle amino acids and other metabolites during nutrient starvation, and to destroy invading pathogens (Cecconi & Levine, 2008; Dooley et al., 2014; Susmita Kaushik & Ana Maria Cuervo, 2018; Mizushima, 2018; Richter et al., 2016). The molecular machinery, the ATG family of proteins, involved in this process was first identified in 1990s in yeast (Ohsumi, 2014), and several conserved mammalian homologues of these core autophagy proteins have since been identified in higher organisms. Basic steps in the autophagy pathway – 'formation' i.e. biogenesis of the phagophore, known previously as an isolation membrane, followed by phagophore membrane 'extension' and 'closure' to form a nascent autophagosome, which subsequently matures through fusion with an endosome to form an amphisome and finally, the lysosome to form an autolysosome, have been conserved across eukaryotic organisms. Interestingly, in yeast, autophagosomes form at a specific site, called the PAS (Phagophore Assembly Site), and no active transport of these structures towards the vacuole (lysosome in mammals) is required. Thus, microtubule- based molecular motors are dispensable for autophagy in yeast. However, in mammalian cells, *de novo* formed autophagosomes need to be actively transported to lysosomal compartment usually located near the center of the cell by microtubule-based motors, particularly, the minus-end directed motor, dynein. Thus, mammalian autophagy also requires contribution of microtubule motor proteins for maturation and completion of the autophagy pathway. Recent studies have shown that post-mitotic mammalian neurons exhibit high levels of basal autophagy. Several neurodevelopmental and degenerative disorders have been shown to arise from defects in the autophagy pathway (Bordi et al., 2016; S. Gowrishankar et al., 2015; Narendra, Tanaka,

Suen, & Youle, 2008). Given the highly polarized morphology of neurons, and sequestration of neuronal lysosomes mostly in the soma, microtubule-based transport of autophagosomes formed distally along the neurites or at growth cones becomes critical for delivery to and degradation of autophagosomes by the lysosomal compartment. However, the molecular machinery involved in autophagosome transport in neuronal and non-neuronal cells, as well as the molecular mechanisms involved in motor recruitment to autophagosomes are only beginning to be understood (X.-T. Cheng, B. Zhou, M.-Y. Lin, Q. Cai, & Z.-H. Sheng, 2015; Fu, Nirschl, & Holzbaur, 2014).

Autophagy has been classified into several sub-types depending on the molecular mechanism and specialized machinery involved, or the type of cellular cargoes being engulfed by the autophagosomes. *Macroautophagy* was first identified as a non-selective process used for bulk degradation of cytoplasmic contents to recycle basic building blocks in cells. This process involves formation of a double-membraned autophagosome around cytoplasmic organelles, cytosolic proteins, or invading microbes, and the subsequent degradation of these autophagic contents by lysosomes, whereupon the amino acids and the breakdown products from the lysosomal lumen are recycled for synthesizing new building blocks for cells to prolong survival during stress or starvation (Yuchen Feng, He, Yao, & Klionsky, 2014). This is the most prevalent form of autophagy, and it is usually triggered by starvation or nutrient deprivation. My work focuses on this branch of autophagy, and molecular details of this process are discussed later in this chapter.

Microautophagy is a non-selective degradative process that involves direct engulfment of soluble autophagic cargoes into the lytic compartment. Microautophagy is employed by cells to maintain membrane homeostasis, organelle size, and to promote survival under nitrogen restriction. During starvation induced microautophagy, extended membranes on the surface of the lysosome rapidly invaginate to form structures called 'autophagic tubes'. These autophagic tubes form a constriction at the neck and eventually undergo scission to capture autophagic cargo for degradation (W.-w. Li, Li, & Bao, 2012). Recent studies have revealed that in steady state, homeostatic cells growing in nutrient-rich conditions, autophagy can be used for selective sequestration of specific subcellular cargoes such as aggregated proteins, pathogens, excess or damaged organelles such as peroxisomes (pexophagy), ER (ER-phagy), mitochondria (mitophagy), Lipid droplets (lipophagy), ribosomes (ribophagy) and a variety of pathogens (xenophagy) (Ding et al., 2010) (Kudchodkar & Levine, 2009) (Singh et al., 2009) (MacIntosh & Bassham, 2011) (Reef et al., 2006) (Khaminets, Behl, & Dikic, 2016; Kraft, Reggiori, & Peter, 2009; Rogov, Dotsch, Johansen, & Kirkin, 2014). These selective autophagy processes occur in nutrient-rich growth conditions, where homeostatic cells get rid of their damaged or excess organelles and invading pathogens. Several studies have now shown that defects in these processes might be causal in many neurodegenerative diseases including Parkinson's disease, Alzheimer's disease and Huntington's disease; cancers, heart and liver diseases and aging (Hübner & Dikic, 2020; Kounakis, Chaniotakis, Markaki, & Tavernarakis, 2019; Um & Yun, 2017).

Chaperone-mediated Autophagy (CMA) is an atypical form of autophagy, where proteins marked for lysosomal degradation contain a degradation tag - the KFERQ motif. These proteins are imported into the lysosomal lumen through a sophisticated mechanism involving chaperones and lysosomal transmembrane channel proteins (S. Kaushik & A. M. Cuervo, 2018). During CMA, the KFERQ-like motif containing proteins are selectively delivered to the lysosome by the HSC70 chaperone complex that includes co-chaperones such as HSC40, HSP90, HSP70-HSP90 organizing protein (HOP) and HSP70-interacting protein (HIP). The HSC70 chaperone complex unfolds the degradative protein substrates before delivering them for lysosomal membrane internalization. The degradative cargoes are internalized into the lysosome via the transmembrane, tetrameric channel LAMP2A (Lysosome-associated membrane protein type 2A), whose C-terminal 12aa (amino acid) cytoplasmic tail is required for the docking of HSC70-substrate complex on the lysosomal membrane. Upon tetramerization of the LAMP2A channel, the substrate is released from the LAMP2A tail and translocated into the lysosomal lumen, where it is degraded (S. Kaushik & A. M. Cuervo, 2018). For the purpose of this study, I have focused on Macroautophagy, hereon referred to as 'Autophagy'.

1.1. Basic Steps in the Autophagy Pathway

Autophagy involves *de novo* in situ formation of a double membraned autophagosome (AP) that encloses cytoplasmic 'cargo' typically marked for degradation through ubiquitination. The formation of this double-membraned autophagosome involves distinct transitional stages namely initiation, elongation and closure, each of which require the contribution of complex multiprotein assemblies.

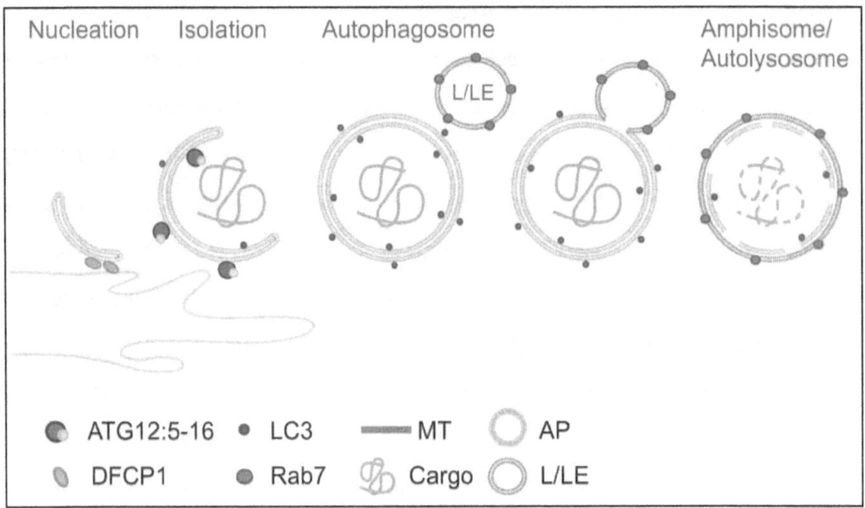

Figure 1. Basic Steps in the Autophagy Pathway.
Autophagosome formation begins at ER nucleation sites decorated with early autophagy proteins such as DFCP1. The isolation membrane (phagophore), labeled with the ATG12:5-16L1 tripartite complex, extends to enclose the autophagic cargo into a fully-formed double-membraned autophagosome, decorated with lipidated LC3 molecules. This nascent autophagosome then fuses with endosomes and lysosomes, to form amphisomes and autophagolysosomes, respectively, where the autophagosomal luminal cargo is degraded by lysosomal proteases.
(*key*: DFCP1: Double FYVE domain containing protein-1, AP: Autophagosome, L/LE: Lysosome/Late Endosome, MT: Microtubule)

Sites of Autophagosome Formation

Initiation

Autophagy machinery is recruited to Phagophore Assembly site (PAS) in yeast or omegasomes in mammals to initiate nucleation of the phagophore (Dikic & Elazar, 2018). Class III PI-3 Kinases play critical roles in regulating membrane trafficking in mammalian cells, including in autophagy **(Lindmo & Stenmark, 2006)**. VPS34, a PI-3 kinase and its accessory protein Beclin-1 (VPS30), are essential for induction of autophagy (Akio Kihara, Takeshi Noda, Naotada Ishihara, & Yoshinori Ohsumi, 2001) (Koyama-Honda, Itakura, Fujiwara, & Mizushima, 2013). Inhibition of PI 3-kinases using small molecule inhibitors such as LY294002, 3-Methyladenine and Wortmannin is sufficient to inhibit autophagy (Blommaart, Krause, Schellens, Vreeling-Sindelarova, & Meijer, 1997). Induction of autophagy results in the activation of the PI3KC3 (Phosphatidylinositol 3-kinase catalytic subunit type 3), which is the catalytic subunit of the PI3K complex. The PI3K complex consists of VPS34, VPS15, Beclin1 and ATG14, which through VPS34-mediated phosphorylation of phosphatidylinositols on the ER membrane, generates PI3P (Phosphatidylinositol 3-phosphate). PI3P in turn serves to accumulate autophagic machinery at that ER site (A. Kihara, T. Noda, N. Ishihara, & Y. Ohsumi, 2001). On mammalian autophagosomes, this lipid is predominantly present in the outer leaflet of the AP membrane, while in yeast it is present on the inner leaflet, suggesting evolutionary differences in the mechanism of AP formation from yeast to mammals (Cheng et al., 2014). The formation of PI3P-rich domains within the ER membrane results in the enrichment of FYVE zinc finger domain containing protein (DFCP1) (Hayashi-Nishino et al., 2009; Ylä-Anttila, Vihinen, Jokitalo, & Eskelinen, 2009). The translocation of DFCP1 from the cytoplasm to the PI3P-rich ER membrane subdomains can also be triggered in response to Beclin1 and VPS34 activity and can be enhanced in the presence of extracellular cues such as nutrient starvation. The domain within DFCP1 spanning amino acids 416-543 is sufficient for its localization to ER, independent of its two FYVE domains (Axe et al., 2008; Cheung, Trinkle-Mulcahy, Cohen, & Lucocq, 2001). In addition to ER membranes, recent studies have shown that APs can be formed from other membranous compartments in a cell including mitochondria, Golgi, ERGIC, plasma membrane, recycling endosomes as well as ER exit sites **(Abada & Elazar, 2014; Lindmo & Stenmark, 2006)**.

Machinery involved in building the autophagosome

As a first step in mammalian autophagosome nucleation, Ulk1 complex: UNC-51 like kinases along with FIP200, ATG13, ATG101, is recruited to the site of phagophore formation (Chan, Longatti, McKnight, & Tooze, 2009). Secondly, PI3K complex (which contains VPS34) is recruited to the phagophore formation site. Following mTOR inhibition or amino acid starvation, Ulk1 is dephosphorylated, which results in its activation. Ulk1 in turn phosphorylates Beclin1 on Serine 14, which activates the PI3K complex (Russell et al., 2013). Finally, the ATG5:12-16L1 complex is recruited to this site to enable elongation of the autophagic membrane. The FIP200 subunit of Ulk1 complex is able to interact with ATG5, which may help in concurrent recruitment of the Ulk1 and ATG5:12-16 complexes to the phagophore (Gammoh, Florey, Overholtzer, & Jiang, 2013) (Nishimura et al., 2013). Mammalian WD repeat domain Phosphoinositide-interacting proteins (WIPIs) have been shown to recruit ATG16L1 to the expanding phagophore. The interaction site within ATG16L1 that binds WIPI2 is distinct from its binding sites for ATG5:12 complex. The ATG5:12-16L1 complex is essential for lipidation of the ATG8 homologues LC3 and GABARAPs onto the growing isolation membrane. The nascent pro-ATG8 proteins (pro-LC3 in mammals) are cleaved by a cysteine protease, ATG4, at their C-terminal domains to expose a glycine residue that is essential for conjugation to PE (Slobodkin & Elazar, 2013). Four different ATG4 homologues exist: ATG4B has been shown to possess widest spectrum of substrates including LC3, ATG4A is more specific to GABARAPs and ATG4C and D show minimal activities (M. Li et al., 2011). The ATG4-processed ATG8s are then activated by an E1- Ubiquitin ligase like enzyme ATG7, and conjugated to PE on the isolation membrane through the activity of another Ub-like enzyme ATG3 (Hamasaki et al., 2013). This process converts nascent free-floating pro-ATG8 homologues (LC3-I) to their membrane-tethered lipidated LC3-II form. ATG7 activates ATG12 and covalently conjugates it to ATG5 through the activity of an E2 -Ub ligase like enzyme ATG10. This ATG5:12 complex then acts like an E3 Ub-ligase complex that serves to facilitate the process of lipidation *in vivo*.

Autophagosome Membrane Elongation and Membrane Sources

The size of the fully-formed double membraned autophagosome is variable, and chiefly dependent on the size of the cargo that is engulfed by the autophagosome. The AP cargoes range from small precursor aminopeptidases (~150nm) to cytoplasmic organelles such as mitochondria and bacteria which can be several micrometers in size (Stolz, Ernst, & Dikic, 2014). After the initiation of autophagosome formation, the phagophore membrane is further expanded by membranes originating from the Golgi, plasma membrane, and recycling endosomes (Lamb, Yoshimori, & Tooze, 2013). The AP membrane expansion can occur through three distinct molecular pathways involving Class III PI3 Kinases, ATG9 and RAB proteins, each of which are located on the Golgi apparatus.

Recent studies have also implicated the ER-Golgi Intermediate compartment (ERGIC) and the COP-II coated vesicular trafficking from Golgi in AP membrane expansion. In particular, ERGIC recruits an early AP nucleation component ATG14, and is both necessary and sufficient for AP formation *in vitro* as well as *in vivo* (Ge, Melville, Zhang, & Schekman, 2013).

ATG9, the only multi-spanning transmembrane ATG family protein, is also critical for AP membrane expansion. ATG9A is enriched in the trans golgi network (TGN) and is exported from the TGN to the periphery of the cell by a multi-subunit adaptor AP-4 (Mattera, Park, De Pace, Guardia, & Bonifacino, 2017). In the absence of AP-4, ATG9 accumulates in the TGN, which in turn leads to a decrease in LC3-I to LC3-II conversion and defects in AP formation, as well as accumulation of autophagic substrates in the cell (De Pace et al., 2018). ATG9 also localizes to the recycling endosome compartment. Syntaxin-18 (SNX18), a membrane remodeling protein, is involved in the membrane tubulation from recycling endosomes, to which ATG9 is recruited and eventually trafficked out of the recycling endosome back to TGN (Soreng et al., 2018). SNX-18 interacts with dynamin-2 for the eventual scission of ATG9-containing tubules which are also positive for ATG16L1. These proteins are then transported to the sites of AP formation which have been shown to be WIPI-2 positive. However, the motors and/or adaptors involved in the tubulation of ATG9-ATG16L1 - positive membranes from REs and their subsequent transport are yet unknown. Under nutrient-rich, steady state conditions, ATG9 is localized to trans golgi network, recycling endosomes and late endosomes, and shows extensive colocalization with TGN-46, Rab7 and Rab9 (Young et al., 2006). Upon nutrient starvation, ATG9 gets redistributed from the TGN onto the late endosomes and vesicular/ tubular compartments near Golgi through the action of BAR protein BAX-interacting Factor-1 (BIF1) (Y. Takahashi et al., 2011). The ATG9-rich vesiculotubular structures are recruited to the phagophore membrane independent of the early ATG proteins such as ULK1 and WIPI2. However, they show a dynamic interaction with phagophore membranes rather than stable structural incorporation into the growing phagophore membrane. Thus, ATG9 compartments may predominantly provide the membrane required for the expansion of the phagophore through transient interactions (Orsi et al., 2012).

At least two Rab proteins associated with the golgi apparatus have been shown to participate in phagophore membrane expansion. GTP-bound active form of Rab1 is essential for the formation of DFCP1-positive nucleation sites on the ER. Another protein, Rab33B has also been shown to directly interact with ATG16L1, and is required for AP membrane expansion, however the exact molecular mechanisms involved remain to be elucidated (Itoh et al., 2008; Mochizuki et al., 2013).

Finally, plasma membrane derived Clathrin-coated vesicles (CCVs) can also provide membrane lipids to the growing phagophore. These newly-formed CCVs become ATG16L1 positive but remain EEA1-negative and grow in size through homotypic fusion facilitated by vesicle membrane associated protein 7 (VAMP-7); and eventually may acquire LC3 (Moreau, Ravikumar, Renna, Puri, & Rubinsztein, 2011). Recycling endosomes may also contribute to AP membrane expansion through TBC1D14 and Syntaxin18 (Knævelsrud et al., 2013; Longatti et al., 2012).

Before the closure of the phagophore membrane, the ATG family proteins dissociate from it, while LC3 and its homologues persist on the fully-formed AP (Klionsky, 2005). Lipidated LC3 is enriched on the inner membrane of the post-closure AP and remains associated until AP degradation by the lysosomal proteases.

After the closure of the newly formed autophagosome, it has to be severed from the membrane source from which it was formed. The machinery involved in this process is only beginning to be identified. Recent evidence suggests that in the final stages of autophagosome formation, ESCRT-III component CHMP2A localizes to the phagophore membrane and mediates the abscission of this membrane. Inhibition of the AAA-ATPase VPS4 which is required for the assembly of functional ESCRT-III complexes also impedes the formation of a fully-formed autophagosome (Yoshinori Takahashi et al., 2018). Further work reveals that Rab5-dependent recruitment of ATG17- vacuolar-sorting protein Snf7 complex to APs results in ESCRT-mediated abscission of AP membrane and subsequent closure (Zhou et al., 2019).

Autophagosome Maturation and Fusion

AP-Lysosome fusion results in the release of the single-membraned autophagic body into the lumen of the lysosome (Dikic & Elazar, 2018). Fully-formed APs first fuse with late endosomes to form amphisomes. Rab11 decorates multivesicular bodies (MVBs) and is necessary for the fusion between APs and MVBs. The v-SNARE VAMP3 is required for AP-MVB fusion but not for the fusion of the resulting amphisome with the lysosome. This latter fusion event requires another v-SNARE protein VAMP7 (Fader, Sánchez, Mestre, & Colombo, 2009). Depleting the ESCRT complex subunits results in decreased autophagic degradation in cells. Especially, mutations in the ESCRT-III subunit CHMP2B also result in the accumulation of ubiquitin-p62-ALFY positive aggregates in cells, suggesting an impaired AP turnover (Filimonenko et al., 2007). APs show two types of fusions with lysosomes: complete fusions and kiss-and-run fusions, where only partial contents of the AP lumen are transferred to the lysosome and both AP and lysosome still persist as two independent organelles (Jahreiss, Menzies, & Rubinsztein, 2008).

Lysosomal distribution in the cytoplasm is sensitive to the cytoplasmic pH. Starvation makes the cytoplasm more alkaline and induces the movement of lysosomes towards the MTOC, i.e. the cell center. For example, upon acidification of cytoplasmic pH to 6.5 due to external stimuli such as acetate treatment, the lysosomes move out towards the very periphery of the cell but return to cell center upon removal of the stimulus (Heuser, 1989; Korolchuk et al., 2011). Thus, upon starvation, the lysosomal compartment is sequestered closer to the MTOC, and newly formed autophagosomes have to be actively transported in the retrograde direction by cytoplasmic dynein towards the lysosomal compartment for their turnover. Thus, starvation might induce dynein-mediated retrograde transport of autophagosomes to increase autophagic flux, a possibility not yet tested experimentally.

Regeneration of the Lysosome from Autophagolysosome

During autophagy, autophagosomes fuse with lysosomes to get their luminal content degraded by the lysosomal proteases. Thus, at the peak of autophagy, APs fuse with lysosomes to form autophagolysosomes (APLs). However, it can be discerned that there is a relative decrease in the total cellular lysosomal pool upon APL formation, and lysosomes have to be reformed after the attenuation of autophagy to restore a steady state pool of cellular lysosomes.

Lysosomal efflux permeases are required for reactivating mTOR kinase and restoring homeostasis. One such efflux permease, named *Spinster*, localizes to the mammalian Lamp1-positive lysosomal membrane. Depletion of spinster from cells results in enlargement of lysosomes, which is even more pronounced upon nutrient starvation. Studies show that in the absence of spinster activity, the autophagic lysosomal reformation in cells in inhibited (Rong et al., 2011).

Lysosomal reformation from APL starts with the initiation of tubulation from the APL membrane. This process of tubulation is mediated by kinesin motor Kif5B. Knockdown of Kif5B and introduction of the motility-defective Kif5B mutant both prevent APL tubulation, suggesting that this particular kinesin motor is essential for APL tubulation. Kif5B directly interacts with PIP2 and is enriched on the PIP2 microdomains on the vesicular membrane. Clathrin and PIP5K1B (Phosphatidylinositol-4-Phosphate 5-Kinase Type 1 Beta) mediate formation of PIP2-rich membrane subdomains on the surface of autolysosomes, which in turn can be bound by Kif5b. In this manner, clathrin may promote kinesin recruitment to APLs and induce tubulation to reform lysosomes (Du et al., 2016). PIP5K1B helps convert PI(4)P to PIP2. Knock down of PIP5K1B indeed results in the reduction of clathrin recruitment to autolysosomes. It is also observed that PIP2 is selectively enriched onto the buds coming off of APL membranes, while PI4P is ubiquitously present on APL membranes. Thus, it appears that PIP5KB1 regulates membrane lipid composition to initiate tubulation and eventual lysosomal reformation.

PIP5K1A (Phosphatidylinositol-4-Phosphate 5-Kinase Type 1 Alpha) also shows a striking localization to the membrane buds i.e. tubulation, and has been shown to be required for the pinching-off of the proto-lysosomes from the APL. Consistent of a role for Clathrin-PIP2 interaction for proto-lysosome fission, depletion of AP2 also results in a block in formation of 'reformation tubules' (which eventually form lysosomes) (Rong et al., 2012). This process of reforming the full complement of lysosomes post-autophagy induction is mTOR-dependent. Upon prolonged starvation and induction of autophagy, lysosomal degradation and recycling of APL cargoes reactivates mTOR and induces the reformation of lysosomes (L. Yu et al., 2010).

It has also been shown that the large GTPase Dynamin 2/ DNM2 is required for the scission of proto-lysosomes from the reformation tubules. Acute or chronic inhibition of Dynamin 2 results in four- to five-fold increase in the size of APLs. When DNM2 inhibition is combined with nutrient starvation, it results in extensive tubulation from the APL compartment. Thus, lysosomal reformation is increased as autophagic flux increases upon nutrient starvation. The APL tubules show a striking localization of DNM2 along their length. When the wild type DNM2 is expressed in cells, this protein localizes to tubules and also shows scission of these tubules into small vesicular structures i.e. proto-lysosomes (Schulze et al., 2013). PI4KIIIbeta (Phosphatidylinositol 4-kinase III beta) with an active kinase domain plays an important role in negative regulation of the APL tubulation. Knockdown of PI4KIIIbeta results in extensive APL tubulation, but no proto-lysosome formation, which in turn causes a loss of lysosomal contents, and failure in maintaining a steady lysosomal pool (Sridhar et al., 2013). mTOR kinase also controls APL membrane tubulation through VPS34 kinase. In nutrient-rich conditions, mTOR kinase directly phosphorylates UVRAG at two serine residues. Phosphorylated UVRAG then binds VPS34 kinase, and activates it, which results in inhibition of APL tubulation. However, upon nutrient starvation and mTOR inhibition, UVRAG and VPS34 kinase complex is deactivated, and tubulation is induced. mTOR, in this way, also promotes scission of newly formed lysosomes by activating UVRAG. This in turn increases the pool of PI3P on APL membrane, the lipid that marks the site of tubule scission by dynamin. This lysosome-localized PI(3)P generated by UVRAG-VPS34 kinase complex has also been shown to be important for the scission of proto-lysosomes (Munson et al., 2015).

A separate study shows two new proteins, Spastizin (SPZ15) and Spatacsin (SPZ11), upon depletion, cause an enlargement of LAMP1 positive compartments in cells. In *wt* cells, prolonged starvation and attenuation of autophagy results in restoration of the size of the autophagolysosomes. However, in Spastizin and Spatacsin depleted cells, the APLs remain enlarged and do not get cleared. Finally, depletion of these proteins results in decreased ALR in cells upon short-term starvation and induction of autophagy (Chang, Lee, & Blackstone, 2014). Consistent with this, Spatacsin knockout in mice results in severe neuronal loss in the cerebral cortex and the cerebellum, mimicking spastic paraplegia-like features. In these mice, Purkinje cells accumulate abnormal large APLs, which in turn has been shown to decrease the regeneration of lysosomes and deplete their steady state pool in these cells (Varga et al., 2015).

Thus, a new set of proteins are being discovered that play an important role in lysosomal reformation from the autophagolysosomes. The process of ALR seems to be critical for maintaining an active steady state pool of lysosomes in cells, and for an efficient turnover of the newly forming autophagosomes. Defects in ALR might be causal for a steady depletion of lysosomal pool in aging and diseased cells, a possibility not yet explored.

1.2. Microtubule-based Transport of Autophagosomes

The Phagophore assembly site (PAS) in yeast is located on the vacuole (similar to the lysosome in mammals) and as such, the APs formed in yeast do not rely on active microtubule-based transport by retrograde motors to reach the vacuole. Unlike in yeast, mammalian autophagosomes can form at diverse sites within the cytoplasm, distant from the lysosomal compartment which is usually located close to the center of the cell. Thus, in mammals, autophagosomes have to be actively transported by the MT-minus end directed motor towards the lysosomal compartment for degradation of the autophagic cargo (Hollenbeck, 1993). Directional transport of autophagosomes is even more critical for the autophagy pathway in highly asymmetric, polarized cells such as neurons (Chklovskii, 2004). The Cathepsin-positive degradative compartment is restricted to neuronal cell body, while the autophagosomes can form at any site within the soma or the neurites, which tend to extend for several hundreds of micrometers *in vivo (Swetha Gowrishankar et al., 2015; Yap, Digilio, McMahon, Garcia, & Winckler, 2018b)*. Thus, especially in neurons, dynein-mediated transport of APs in axons and potentially, dynein and kinesin-mediated transport of APs in dendrites (given the mixed polarity of microtubules) is critical for AP turnover (Millecamps & Julien, 2013). Several recent studies in mammalian neurons have shown that cytoplasmic dynein adaptor proteins have the ability to directly bind Rab7 on the late, post-LE fusion autophagolysosomes, as well as to LC3 on nascent autophagosomes. Thus, both APs and APLs can be actively transported along microtubules towards the lysosomal compartment for their degradation. One study has shown that JIP1, a motor adaptor protein, recruits both the anterograde motor kinesin and the dynein accessory protein, dynactin. JIP1 is required for the retrograde transport of a subset of APs in axons (Fu et al., 2014). Phosphorylation of a serine residue within JIP1 acts as a switch between retrograde and anterograde transport, thus allowing bidirectional transport of autophagosomes in neurons. In addition, LC3 binding to JIP1 through its conserved LIR (LC3-Interacting Region) motif also acts as a regulatory mechanism to prevent kinesin binding to JIP1, and to ensure a robust retrograde transport of APs within the axon.

Another study on a motor adaptor protein SNAPIN proposes a different mechanism for AP transport. SNAPIN associates with the late endosomal membranes and recruits dynein to LEs. Upon LE fusion with APs, the SNAPIN bound motor is in turn acquired by the newly formed autophagolysosome (APL) which then starts moving retrogradely towards the cell body of the neuron (X. T. Cheng, B. Zhou, M. Y. Lin, Q. Cai, & Z. H. Sheng, 2015). Furthermore, in keeping with this model, inhibition of AP-LE fusion by knocking down Syntaxin-17, results in decreased acquisition of dynein by APs, inhibition of their retrograde transport and consequently, accumulation of APs in the distal axon of DRG neurons.

Distally formed neuronal autophagosomes are also involved in the AP-2 mediated retrograde transport of BDNF-activated TrkB receptors to neuronal soma (Kononenko et al., 2017). This function is important for neuronal survival, and for maintaining axonal integrity and preventing axonal degeneration.

There is also a possibility that the components of the core autophagy machinery might be actively transported by microtubule motor- based transport to subcellular sites of AP formation. However, the specific motors and adaptors involved in this process need to be investigated.

1.3. Regulation of Autophagy Pathway in Mammals

The process of autophagy is highly regulated in mammalian cells. During homeostatic conditions, autophagy is suppressed, however basal low levels of autophagy are detected in cells.

Upon nutrient deprivation or acute stress, autophagy is upregulated which triggers formation of new autophagosomes and degradation of ubiquitinated substrates. The autophagy pathway is controlled by a number of different cellular signaling pathways, which sense and respond to cell stress and nutrient availability, amongst other environmental cues. The major autophagy regulator is mTOR (mechanistic target of rapamycin) kinase, which negatively regulates autophagy during homeostatic conditions.

<u>mechanistic Target of Rapamycin (mTOR)</u>

mTOR kinase serves a critical role in sensing nutrient deprivation and other forms of stress. Inhibition of mTOR activates autophagy and prolongs cell survival. However, how mTOR regulates neuronal autophagy remains a critical, but poorly understood question (J. Kim, M. Kundu, B. Viollet, & K.-L. Guan, 2011; Maday & Holzbaur, 2016; Robert A. Saxton & Sabatini, 2017).

Several studies in non-neuronal cells have addressed the mechanism of mTOR control of autophagy pathway. During steady-state growth conditions, mTOR constitutively inhibits autophagy by phosphorylating several key protein kinases required to initiate autophagosome formation. mTOR phosphorylates ULK1 kinase, which prevents its association with autophagy proteins FIP200, ATG101 and ATG13. This in turn inhibits phagophore initiation and blocks AP formation (J. Kim, M. Kundu, B. Viollet, & K. L. Guan, 2011). In nutrient-rich conditions, an active mTOR kinase also phosphorylates TF-EB, a transcription factor required for the transcription of lysosomal biogenesis and autophagy-related genes (A. Roczniak-Ferguson et al., 2012). During nutrient-rich conditions, mTOR phosphorylation of TF-EB prevents its nuclear translocation, thus blocking the transcription of TF-EB downstream target genes. Inhibition of mTOR kinase or nutrient starvation results in TF-EB translocation to the lysosomal membrane where it co-localizes with mTOR, and finally to the nucleus. In steady state cells, mTOR phosphorylates TF-EB at Serine 211, which induces TF-EB binding to 14-3-3 proteins and its retention in the cytoplasm. Inhibition of mTOR signaling upon starvation results in decreased TF-EB phosphorylation and loss of 14-3-3 mediated retention of the transcription factor in the cytoplasm, leading to its nuclear translocation (Agnes Roczniak-Ferguson et al., 2012). This is an obligate step to upregulate autophagy-specific genes. Consistent with this finding, depletion of TF-EB prevents the upregulation of lysosome biogenesis and autophagy genes in cells even upon nutrient starvation (Settembre et al., 2012). Upon amino acid feeding, and mTOR kinase reactivation, mTOR phosphorylates nuclear TF-EB on Serine residues 138 and 142 in close proximity of a nuclear export signal, which induces its CRM1-mediated export from the nucleus.

mTOR plays a critical role in nutrient sensing in mammalian cells. Rag GTPases (RagA, B, C and D) are Ras family proteins that control mTOR signaling pathway *via* amino acid sensing. These proteins regulate mTOR localization to the Rab7-positive compartment in cells, which brings mTOR spatially closer to its activator, Rheb (Sancak et al., 2008). Constitutively active RagB expression induces hyperactivation of mTOR kinase and its constitutive localization at the Rab7-positive membranes, irrespective of amino acid levels in cells, thus making mTOR resistant to amino acid signaling (Sancak et al., 2008). Ragulator, which consists of a multi-subunit protein complex, including p14, p18, MP1, C9orf52 and HBXIP, promotes GTP nucleotide loading onto the RagA-B homodimeric complex, thus serving as a GEF for RagA and RagB (L. Bar-Peled, Lawrence D. Schweitzer, R. Zoncu, & David M. Sabatini, 2012). This interaction between Rag GTPases and Ragulator becomes stronger upon amino acid starvation and weakens upon refeeding, where the Ragulator shows preferential binding to nucleotide-free Rag proteins, consistent with its role as a GEF. Ragulator is critical for the recruitment of Rag GTPases to lysosomal membranes, which then recruit mTOR to lysosomes in an amino acid -dependent manner. mTOR kinase undertakes nutrient sensing in conjunction with several multiprotein complexes, such as GATOR1, GATOR2, Sestrins, Folliculin-FNIP and KICSTOR, to regulate the amino acid dependent mTOR signaling pathway (Bar-Peled et al., 2013; L. Bar-Peled, L. D. Schweitzer, R. Zoncu, & D. M. Sabatini, 2012; Chantranupong et al., 2016; Chantranupong et al., 2014; R. A. Saxton, Chantranupong, Knockenhauer, Schwartz, & Sabatini, 2016; Tsun et al.,

2013; R. L. Wolfson et al., 2017; Wyant et al., 2017). Recent works have also identified specific proteins that bind particular nutrients and sense their abundance in the cellular cytoplasm. For example, SLC38A9 senses lysosomal arginine concentration (S. Wang et al., 2015), while CASTOR1 and Sestrin 2 sense cytosolic arginine and leucine, respectively (Chantranupong et al., 2016; Rachel L. Wolfson et al., 2016).

<u>AMP-activated protein kinase (AMPK)</u>

AMPK is activated in cells upon glucose depletion or ATP reduction. AMPK inhibits mTOR kinase activity through phosphorylation. An activated AMPK phosphorylates the TSC2 subunit of the Tuberous Sclerosis Complex. This phosphorylation results in the activation of TSC2. The activated TSC2 functions as a GTPase-activating protein that in conjunction with TSC1, converts Rheb-GTPase to its inactive, GDP-bound form. Loss of Rheb-GTP from the lysosomal membrane decreases mTOR kinase function (Luo, Saha, Xiang, & Ruderman, 2005) and activates autophagy. A stress stimulus like glucose starvation results in AMPK phosphorylation of ULK1 kinase, which promotes its association with FIP200, ATG13 and ATG101 to initiate AP formation (Kim et al., 2011b). Thus, AMPK senses energy levels in cellular cytoplasm and acts as a positive regulator of autophagy in mammals. The relative activities of AMPK and mTOR kinases regulate the degree of autophagy induction in cells.

AMPK depleted cells also show a defect in the maturation of APs to APLs, where LC3-positive nascent APs accumulate in these cells but show a decreased co-localization with LAMP-1 positive late endosomes. However, the exact mechanism by which AMPK regulates AP maturation is not yet completely understood (Jang et al., 2018).

<u>Protein Kinase A (PKA)</u>

cAMP-dependent Protein Kinase A (PKA) has emerged as the negative regulator of autophagy in yeast (Yorimitsu, Zaman, Broach, & Klionsky, 2007). In nutrient-rich conditions, small GTPases Ras1 and Ras2 increase Adenylyl Cyclase (AC) mediated production of cyclic AMP (cAMP). cAMP in turn activates PKA which directly phosphorylates both ATG1 (Budovskaya,

Stephan, Deminoff, & Herman, 2005) and ATG13 proteins (Stephan, Yeh, Ramachandran, Deminoff, & Herman, 2009b). Interestingly, the residues phosphorylated by PKA in ATG13 are distinct from the residues phosphorylated by mTOR kinase. PKA phosphorylation of ATG13 prevents its recruitment to PAS in yeast, which is an upstream event to ATG1 recruitment to PAS. This abolishes ATG1 arrival at the PAS, effectively preventing AP formation. In keeping with these observations, inhibition of PKA results in robust induction of autophagy in yeast (Stephan, Yeh, Ramachandran, Deminoff, & Herman, 2009a). In mammalian cells, PKA has been shown to directly phosphorylate ATG16L1 alpha and beta subunits, which marks these proteins for degradation. Thus, in mammals, increased PKA activity results in suppression of autophagy, which is consistent with the above-mentioned studies in yeast. Upon inhibition of PKA, cellular levels of ATG16L1 increase and in turn autophagosome biogenesis is induced (Yorimitsu et al., 2007; Zhao et al., 2019). Another study in mammals shows that the regulatory subunit I of PKA (RI alpha) localizes to LC3-positive APs and Rab7- positive LEs, and co-immunoprecipitates with mTOR kinase. Inhibition of RI alpha which results in over-activation of PKA increases phosphorylation of mTOR kinase which in turn suppresses autophagy (Mavrakis, Lippincott-Schwartz, Stratakis, & Bossis, 2006). Thus, in mammalian cells, PKA may control autophagy directly by phosphorylating a key biogenesis protein, or indirectly through upregulation of mTOR kinase function, however the molecular mechanisms underlying PKA control of mammalian autophagy remain to be fully understood.

<u>Insulin Signaling</u>

Insulin signaling pathway has been shown to negatively regulate autophagy. Insulin signaling downregulates autophagy by reducing cellular expression of LC3-II and by increasing AKT and S6 kinase phosphorylation (Ribeiro, López de Figueroa, Blanco, Mendes, & Caramés, 2016). There is increasing evidence that mTOR overactivation results in insulin resistance in cells (Guillén & Benito, 2018). AKT (Protein kinase B) is a serine/threonine kinase which when activated, inhibits negative regulators of mTOR kinase, such as TSC1/2. This results in activation of mTOR kinase and suppression of autophagy. The AKT signaling pathway is triggered by the binding of insulin and insulin-like growth factor 1 (IGF-1) to the transmembrane IGF-1 receptor, which in turn activates insulin receptor substrate 1 (IRS-1) (de Mello et al., 2019). IRS-1 activates Phopshoinositol-3 kinase (PI3K) and generates phosphatidylinositol (3,4,5)-trisphosphate (PIP$_3$) (Guillén & Benito, 2018). PIP$_3$ has been shown to directly activate mTORC2 complex, which results in suppression of autophagy (Gan, Wang, Su, & Wu, 2011) (Dibble & Cantley, 2015). Interestingly, while IRS-1 is involved in AKT mediated activation of mTOR signaling and suppression of autophagy, it has also been shown that overactivation of autophagy by insulin receptor substrate-2 (IRS-2) results in clearance of cytoplasmic protein aggregates (de Mello et al., 2019; S. Yamamoto et al., 2018). Independent of mTOR kinase, AKT and S6 kinase activity, IRS-2 is able to induce macroautophagy in cells via beclin-1 and VPS34. Thus, an accumulation of cytotoxic protein aggregates in cells might result in mTOR-independent activation of macroautophagy and lysosomal degradation of these aggregates (A. Yamamoto, Cremona, & Rothman, 2006).

1.4. Autophagy in the Nervous System

Neurons are post-mitotic and survive during an individual's entire life-time. Autophagic degradation of cellular waste or damaged and misfolded proteins is especially important for maintaining neuronal homeostasis. However, functions of autophagy pathway in the brain remain poorly understood. All key autophagy proteins are expressed in the central nervous system however, autophagy is only recently getting implicated into several neurodevelopmental and neurodegenerative diseases. The findings so far show that autophagy may play critical roles in not only clearing protein aggregates, but also in basic neurodevelopmental processes, as well as neuronal functioning.

Knockout of an essential autophagy protein, ATG5, in neural cells results in diffused cytoplasmic accumulation of ubiquitinated proteins. This causes increased neuronal death in the cerebral cortex and the cerebellum, and progressive degeneration during early embryonic and perinatal brain development. This finding suggests that normal autophagy is important for homeostasis and impediment of this pathway results in neurodegeneration. These ATG5 knockout mice show progressive increase in ubiquitinated cytoplasmic inclusions with age, another feature usually associated with several neurodegenerative disorders (Hara et al., 2006). Further work shows that an autophagic cargo adaptor protein, p62/ Sequestosome-1 is essential for the formation of the ubiquitinated aggregates or inclusions in neurons and is also a part of the aggregate-containing inclusions. Depletion of p62 inhibits the formation of ubiquitinated aggregates in neuronal cells and also prevents the recruitment of LC3 to growing autophagosomes under nutrient starvation (Bjørkøy et al., 2005). However, the ablation of p62 gene in neurons does not lead to either increase or decrease in the severity of pathology caused by autophagy deficits (Ma, Attarwala, & Xie, 2019). P62 undergoes oligomerization that stabilizes its interaction with LC3B and linear ubiquitin present on the surfaces of target proteins to recognize cargo and help initiate isolation membrane formation around it (Wurzer et al., 2015). Altogether, p62 seems to play an important role in autophagic cargo recognition in neurons and links polyubiquitinated protein aggregates to autophagic machinery.

Autophagy in aggregate-prone diseases

Several neurodegenerative diseases are characterized by an accumulation of cytotoxic protein aggregates that result in progressive neuronal degeneration and death. Emerging evidence suggests that defects in the autophagy-lysosomal pathways may result in cytotoxic neuronal aggregates and modulating the autophagy pathway might hold therapeutic promise for treatment of these disorders.

Alzheimer's Disease is characterized by accumulation of amyloid beta protein in extracellular plaques and cytoplasmic microtubule-binding protein Tau neurofibrillary tangles in cells of the brain. In Alzheimer's disease patients, autophagosomes accumulate in the neurons of frontoparietal cortex (Ralph A. Nixon et al., 2005). This accumulation of autophagosomes has been attributed to reduced function of the lysosomal protease Cathepsin D, which is necessary for the degradation of autophagic cargoes in cells (Papassotiropoulos et al., 2000). Neuronal autophagosomes contain amyloid beta peptides on their membranes and induction of autophagy results in increased amyloid beta secretion (Ntsapi, Lumkwana, Swart, du Toit, & Loos, 2018). Inhibition of autophagy by genetic knockout of an essential autophagy protein ATG7 in Amyloid Precursor Protein (APP) mouse models shows a striking decrease in amyloid beta secretion (Nilsson et al., 2013), which is restored upon autophagy restoration (Nilsson & Saido, 2014). Consistent with this, induction of autophagy by mTOR inhibition induces clearance of amyloid beta aggregates in neurons (Caccamo, Majumder, Richardson, Strong, & Oddo, 2010). Overall, defects in autophagic turnover seem to contribute to amyloid beta accumulation in Alzheimer's disease, however the exact molecular mechanisms that go awry in AD remain to be understood.

Brains of _Parkinson's Disease_ patients show an accumulation of alpha-Synuclein rich fibrillary aggregates called _Lewy bodies_ in neuronal cell bodies as well as neurites, which results in a progressive loss of dopaminergic neurons (Dickson et al., 2009). PD affected neurons also show an accumulation of amyloid beta aggregates and hyper-phosphorylated tau, which seems to stem from defects in autophagy-lysosomal pathways in these neurons (Moors et al., 2016; Y. Wang et al., 2009). PD patient brains show elevated levels of lipidated LC3 (LC3-II) in the Substantia Nigra within the striatum. LC3 tends to colocalize with alpha-Synuclein clusters in this region of the brain (Dehay et al., 2010). Furthermore, LAMP2A, the transmembrane protein involved in chaperone- mediated autophagy (CMA) shows decreased expression in PD patient brains (Alvarez-Erviti et al., 2010; Murphy et al., 2015). Finally, there is increasing evidence of accumulation of damaged mitochondria in PD brain tissue, which indicates an impairment of mitophagy pathway in this disease (Zhu, Guo, Shelburne, Watkins, & Chu, 2003). Despite these findings, the exact molecular mechanisms that go awry in PD remain largely unclear.

Huntington's Disease is characterized by the accumulation of polyglutamine-repeat expansion containing mutant huntingtin (_htt_) protein in neuronal cytoplasm. Interestingly, the accumulation of PolyQ-htt aggregates in neurons seems to induce increased autophagosome formation, however, these autophagosomes often do not contain cargo. Thus, there appears to be a defect in autophagic cargo recognition by autophagy receptors such as p62, which results in inefficient autophagosome formation, and thus a failure in clearance of toxic protein aggregates (Martinez-Vicente et al., 2010). Wild-type *Htt* protein is involved in membrane curvature and autophagosome formation in steady state healthy cells. It is also responsive to ER stress and induces autophagy during ER stress. PolyQ-*Htt* also contains functional KFERQ-like motifs which are necessary for LAMP2A-mediated CMA (Qi et al., 2012). It is speculated that PolyQ-*htt* might not be efficiently translocated across the lysosomal membrane, leading to a backlog in LAMP2A-mediated CMA and an accumulation of PolyQ-*htt* in the cytoplasm. However, the exact molecular functions of *htt* in neurons and the mechanisms involved in PolyQ-*htt* aggregate clearance remain poorly understood.

Autophagy in Neurodevelopment

Post-synaptic Roles for Autophagy

Emerging evidence suggests that autophagy is important for several basic neuronal processes, including synaptic turnover, ion channel turnover and organelle recycling. Upon nutrient deprivation, autophagy is regulated differentially in different brain regions. Specifically, the levels of total and cleaved LC3 go down in the cortex and hippocampus but increase in the hypothalamus and remain unchanged in the cerebellum (V. Nikoletopoulou, K. Sidiropoulou, E. Kallergi, Y. Dalezios, & N. Tavernarakis, 2017). In the forebrain, this suppression of autophagy is correlated with increased BDNF (Brain-derived neurotrophic factor) and TrkB (BDNF transporter) expression, and an inhibition of autophagic flux that results in increased levels of p62 in neurons. On the contrary, depletion of BDNF results in increased autophagic activity in the cortex and the hippocampus, an increased autophagic flux and decreased p62 levels. This downregulation by BDNF is critical for neuronal survival. It suppresses autophagy to increase dendritic spine density which enhances learning and memory. Finally, this BDNF-mediated autophagy might also play a role in directly controlling degradation of key synaptic proteins like SHANK3, PSD95 and PICK1.

Induction of LTD (long-term depression) in hippocampal neurons results in an increased LC3-II positive AP formation in neuronal dendrites. These autophagosomes retrieve AMPA receptor subunits from dendritic spines in response to LTD (Shehata, Matsumura, Okubo-Suzuki, Ohkawa, & Inokuchi, 2012). Similarly, NMDAR and mGluR- mediated long term depression which results in elimination of dendritic spines, includes local formation of autophagosomes, which retrieve synaptic and cytoskeletal components from depressed spines (Kallergi et al., 2020). Induction of LTD in dendrites results in an upregulation of key early autophagy proteins such as Ulk1, ATG101 and ATG13 which are needed for AP biogenesis through local translation. Interestingly, inhibition of autophagosome biogenesis impairs LTD-mediated spine pruning. The autophagic vesicles formed in dendrites upon LTD contain a variety of synaptic proteins, many of which have been implicated in ASD. In accordance with this observation, suppression of autophagy-mediated LTD results in cognitive flexibility deficits, further highlighting the need for autophagy in regulating higher order brain functions. Loss of presynaptic GABA and Acetylcholine inputs results in a loss of GABA receptors in post-synaptic spines. These diffuse GABA receptors are endocytosed and eventually delivered to autophagosomes which transport them for lysosomal degradation (Rowland, Richmond, Olsen, Hall, & Bamber, 2006).

mTOR-mediated autophagy also seems to play a critical role in dendritic spine pruning, a process dysregulated in ASD brains. mTOR overactivation in ASD results in suppression of autophagy which in turn results in decreased spine pruning and an increased spine density in ASD forebrains (especially neurons of layer III). Inhibition of mTOR kinase activity by rapamycin treatment, however, restores dendritic spine pruning and normal spine density. The behavioral deficits exhibited by ASD mice are also corrected upon rapamycin-mediated mTOR inhibition and restoration of autophagy. Thus, normal autophagic functioning is important for basic neuronal function and cognition (Guomei Tang et al., 2014).

Autophagy also plays distinct roles in regulating function and behavior of different neuronal subtypes in the striatum. Striatal direct spiny projection neurons (dSPNs) and indirect spiny projection neurons (iSPNs) are both affected differentially upon autophagy suppression by genetic knockout of ATG7. The dendritic architecture and synaptic inputs of dSPNs are affected upon autophagy suppression, while inward rectifying potassium channel Kir2 turnover is affected in iSPNs resulting in intrinsic hyper-excitability (Lieberman et al., 2020). Thus, autophagy may play very distinct functions in different cell types, an avenue that needs to be explored further. In conclusion, autophagy might directly remodel synapses in response to trophic factor signaling, or alter ion channel turnover to alter excitability of neurons as well as dendritic arborization and spine density; however, further mechanistic insights into autophagic contributions to these fundamental neuronal processes are needed to understand disease pathology and to devise therapeutic strategies.

Pre-synaptic Roles for Autophagy

Autophagy is involved in the degradation of presynaptic vesicles. Small GTPase Rab26 has been shown to cluster synaptic vesicles in neurites as well as in the cell body, to target them for degradation (Binotti et al., 2015). These Rab26-positive vesicles are enriched in autophagosome membrane markers as well. Furthermore, mTOR inhibition in dopaminergic neurons results in autophagosome formation, reduction in axon terminal volume, reduction in the number of synaptic vesicles, as well as increased dopamine release, suggesting that mTOR inhibition results in induction of autophagy that, on a fast temporal scale, can alter axonal morphology, pre-synaptic composition and dopamine release patterns (Hernandez et al., 2012). A Pleckstrin Homology Domain containing protein, Plekhg5, has also been recently implicated in pre-synaptic autophagy (Lüningschrör et al., 2017). Depletion of Plekhg5 results in an inhibition of axon growth, as well as an accumulation of synaptic vesicles in axon terminals. Plekhg5 depletion results in increased axon size, an accumulation of synaptophysin-positive vesicles at axon terminals and loss of axonal structural integrity. This phenomenon is related to the function of Plekhg5 as a Rab26 GEF, which also associates with synaptic autophagosomes. Depletion of Plekhg5 results in impairment of active Rab26 in neurons, which when rescued by Rab26 overexpression, restores normal axonal structure and function (Kim et al., 2013). Prolonged neuronal activity also triggers formation of synaptic LC3-positive autophagosomes, possibly as a mechanism to cope with synaptic stress. Endophillin A, a pre-synaptic protein, recruits ATG3 to synaptic autophagic membranes, and depletion of endophillin results in inhibition of autophagosome formation upon nerve stimulation, and overall defects in synaptic morphology and axonal degeneration over time. Interestingly, it has been shown that the Parkinson's disease kinase LRRK2 activates endophillin A through phosphorylation, which then induces synaptic autophagosome formation (Soukup et al., 2016). Thus, presynaptic autophagy is necessary for proper neuronal functioning and survival.

1.5. Autophagosome Transport: Outstanding Questions

Although the basic stages in autophagosome biogenesis and maturation have been extensively characterized in yeast and mammalian cells, they remain to be fully elucidated in neurons. Neuronal autophagosome behavior, including microtubule-based motility, is also of particular interest given the extreme asymmetry of these cells and the concentration of degradative organelles in the soma (S. Gowrishankar et al., 2015; Yap, Digilio, McMahon, Garcia, & Winckler, 2018a). Neuronal autophagosomes are transported along microtubules, though the stage of AP biogenesis at which motors are first acquired and the molecular mechanisms for regulation of transport, are incompletely understood (Joungmok Kim et al., 2011; Maday & Holzbaur, 2016; Robert A. Saxton & Sabatini, 2017; Takei & Nawa, 2014). Snapin and JIP1 have been implicated in microtubule motor protein recruitment to late autophagosomes (after fusion with Rab7-positive vesicles) in dorsal root ganglion (DRG) neurons. However, neither of these proteins was implicated in direct recruitment of the retrograde motor protein cytoplasmic dynein to autophagosomes (X. T. Cheng et al., 2015; Fu et al., 2014).

1.5. Cytoplasmic Dynein: Microtubule-based Retrograde Motor Essential for Autophagosome Transport

In mammalian cells, microtubule- associated motors play a critical role in controlling the spatial distribution of subcellular organelles. Microtubule minus end- directed transport is under control of the highly-conserved multi-subunit cytoplasmic dynein 1 (from here on, dynein) motor complex (Paschal & Vallee, 1987; R.B. Vallee, J.S. Wall, B.M. Paschal, & H.S. Shpetner, 1988; R. B. Vallee, J. S. Wall, B. M. Paschal, & H. S. Shpetner, 1988). Dynein interacts with a broad range of subcellular cargoes and is responsible for their active transport from cell periphery towards the MT minus ends, which are usually located near the cell center, closer to the nucleus in fibroblasts. Structurally, dynein consists of two motor domain-containing heavy chains (HCs), two each of intermediate (ICs) and light intermediate chains (LICs), and three pairs of light chains (LCs), altogether constituting a ~ 1.6 MDa multi-subunit motor complex. The heavy chains which encode dynein motor domain, are responsible for the force production activity of the motor, while the ICs, LICs and LCs interact with a variety of individual cargoes, as well as a class of proteins called dynein 'regulatory factors'. These mainly include Dynactin, NudE, NudEL and LIS1. One well-characterized dynein regulatory factor is dynactin, which controls the processivity and diffusivity of dynein. Dynactin itself is a ~1.2 MDa multiprotein complex comprising of at least 11 different subunits and has been shown to be essential for dynein function in mammalian cells. On the contrary, LIS1, which was first identified as the causal gene for human neurodevelopmental disorder *lissencephaly*, also functions as a dynein regulatory factor and is specifically required for "high-load" cargo transport (McKenney, Vershinin, Kunwar, Vallee, & Gross, 2010; Yi et al., 2011). LIS1 and dynein form a tripartite complex with yet another regulatory protein encoded by the *microcephaly* gene Nde1 or its paralogue, Ndel1 (NudE and NudEL) (Efimov & Morris, 2000; Y. Feng et al., 2000; Niethammer et al., 2000; Sasaki et al., 2000; Stehman, Chen, McKenney, & Vallee, 2007). This Lis1-NudE-dynein complex is implicated in transport of large cargoes such as nuclei (Shu et al., 2004; Tsai, Bremner, & Vallee, 2007; Tsai, Chen, Kriegstein, & Vallee, 2005;

Tsai, Lian, Kemal, Kriegstein, & Vallee, 2010) and large vesicular organelles of the endolysosomal compartment (Pandey & Smith, 2011; Reddy et al., 2016; Yi et al., 2011). Lis1 has also been shown to promote dynein and dynactin recruitment to mRNA localization complexes to control mRNA distribution in cells (Dix et al., 2013). Interestingly, LIS1 is involved in the motility of dynein-dynactin complexes along microtubules (Egan, Tan, & Reck-Peterson, 2012) and in increasing the duration of individual dynein motor's attachment to microtubules even during cycles of ATP hydrolysis which would detach the motor from the microtubule by blocking dynein's mechanochemical cycle (Huang, Roberts, Leschziner, & Reck-Peterson, 2012; Toropova et al., 2014).

Dynein is recruited to a variety of subcellular cargoes. An interesting question then becomes as to how this highly conserved multi-subunit motor gets specifically recruited to a sub-class of cellular organelles in a spatially and temporally controlled manner. The specificity of dynein recruitment to its diverse classes of cargoes has been evolutionarily achieved through a new, recently identified family of extended α-helical coiled-coil domain containing dynein 'adaptor' proteins. Most of these 'adaptor' proteins exhibit an ability to bind to both dynein and Rab GTPases, and are thus named RDDs (Rab-dynein-dynactin interacting proteins) (Scherer, Yi, & Vallee, 2014). This family of proteins consists of several isoforms of Hook, BicD, FIP, Spindly and now, RILP (Hoogenraad & Akhmanova, 2016; McKenney, Huynh, Tanenbaum, Bhabha, & Vale, 2014). Interestingly, these proteins show very little primary sequence conservation (Gama et al., 2017; Lee et al., 2018), however they clearly share a common domain organization which is functionally conserved, suggesting a similarity in their cellular roles. One such well-characterized RDD protein is called BicD2, which has been shown to recruit dynein and dynactin to Rab6-GTP positive Golgi vesicles (Hoogenraad et al., 2001; Huynh & Vale, 2017; Matanis et al., 2002; McKenney et al., 2014), as well as to other cellular cargoes including the nucleus (Bolhy et al.,

2011; Hu et al., 2013; Splinter et al., 2010). Interestingly, further work has shown that BicD2's role is not limited to passive recruitment of the motor to cellular cargoes but in addition, BicD2 increases the observed processivity of the dynein-dynactin complex (McKenney et al., 2014; Schlager, Hoang, Urnavicius, Bullock, & Carter, 2014) as well as dynein force production (Belyy et al., 2016). Thus, BicD2 in addition to recruiting dynein to cargoes, also serves to alter the biophysical properties of the motor complex itself. LIS1, a dynein regulator implicated in *lissencephaly*, has not been detected as a stable accessory protein of the BicD2-dynein complex *in vivo* (McKenney et al., 2014), but associates with the pre-assembled dynein complex and clearly stimulates its velocity when added *in vitro* (Gutierrez, Ackermann, Vershinin, & McKenney, 2017). This sets precedence for similar motor regulatory functions for other members of the RDD family of proteins and increasing evidence suggests that indeed these adaptor proteins may uniquely alter dynein biophysical behavior in order to expand the range of functions that dynein performs in the cell.

1.7. The Function and Regulation of Dynein Motor Complex

Molecular Interactions between the Tri-partite Complex of Dynein-Dynactin-Adaptor protein

Dynactin, a twenty-three subunit ~1.0MDa complex, in the presence of a dynein adaptor protein such as BicD2, interacts with dynein which itself is a 1.6 MDa multi-subunit complex. Dynactin consists of an Arp1 filament consisting of 8 Arp subunits and one β - actin subunit. The

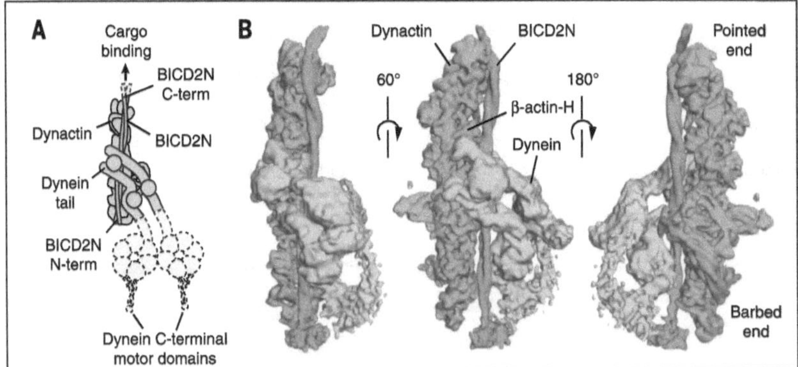

'pointed' and the 'barbed' ends of dynactin Arp1 filament are capped by complexes of proteins: The pointed end is capped by p25/p27/p62 dimers that cap Arp11, while the barbed end is capped by CapZ alpha-beta. The 'shoulder' of the dynactin complex consists of two molecules each of the long extended p150^{Glued} and a dimer of p24 and a tetramer of p50 (dynamitin). The length of

Figure 2. Organization of the Dynein-Dynactin-Adaptor complex.
(**A** and **B**) The N-terminal coiled-coil domain of BicD2 runs parallel along the Arp1 filament of the dynactin supercomplex, and recruits the N-terminal tails of dynein heavy chain to form a stable tripartite complex. A translational symmetry between the adaptor and dynein heavy chain tails organizes the motor domains parallel to each other, allowing high processivity by the motor complex. (Adapted from Urnavicius et. al. 2015)

the Arp1 filament is considered critical in forming a translational symmetry between the N-terminus of the BicD2 adaptor, and the N-terminal tails of DHCs, which allows for the parallel orientation of the two motor domains in DHCs, which facilitates high processivity by the motor (Urnavicius et al., 2015). The Arp1 filament directly binds to DHC tails, and this interaction is further stabilized by the N-terminal ~270 aa coiled coil of BicD2 N-terminus running along the length of Arp1 filament. Since this three-way binding is required for forming a processive dynein motor complex, this arrangement ensures that dynein is not stochastically bound to its cargoes, but can only selectively get recruited to cargoes through the adaptors.

Adaptor Regulation of Dynein Force Production

The interaction of Adaptor N-terminus (e.g. BicD2 N-terminus, BicD2N) with dynactin and dynein to form a stable triple-complex also enhances force production by human dynein, as exhibited *in vitro (Belyy et al., 2016)*. Human dynein shows very low processivity (50nm/ sec), which is almost similar to diffusivity when bound to MTs alone. The stall force for such free dynein molecule is ~2pN. In contrast, human kinesin exhibits much higher stall force of ~ 6pN. Thus, it has been previously proposed that clustering of several dynein motors might be required for competing against the MT plus-end directed kinesin motors. However, the *in vivo* relative ratios of dynein *vs* kinesin motors on individual axonal vesicles suggest that they are indeed almost equivalent (1.5 dynein: 1 kinesin). This suggests that while dynein alone might not be able to resist plus-end directed forces by kinesin, there exists a mechanism where dynactin and BicD2N binding alters dynein force production. Indeed, recent work shows that dynein alone, as well in complex with dynactin, exhibits low stall forces of ~2pN, however, when in a triple complex with BicD2N, it exhibits a substantially high stall force of ~4.5pN, which, in *in vitro* tug-of-war studies, appears sufficient to resist ~6pN force generated by a single kinesin motor in the opposite direction. While dynein alone gets passively transported by a kinesin in the plus-end direction during tug-of-war, addition of dynactin and BicD2N prevents the passive plus-end movement, instead a higher percentage of molecules exhibit at least 20-fold increase in minus-end directed velocity. Thus, adaptor proteins increase dynein force production and enable the retrograde motor to resist kinesin driven plus-end directed motility.

Adaptor Recruitment of Two Dyneins to Dynactin for Increased Force and Velocity Production

Dynein velocity and force production differs between the dynein-dynactin complexes formed with different adaptors. This raises two possibilities: (i) An adaptor may recruit more than one dynein motors to dynactin to increase force production and processivity, or (ii) Every adaptor may modulate a single, bound dynein's activity differently based on its contacts with DHCs and within the dynactin Arp1 filament. Indeed, it has been shown that unlike BicD2, BicDR1 and HOOK3 recruit two dynein motors per dynactin (Urnavicius et al., 2018). While Dynein-dynactin-BicD2 produce stall force of ~3.7pN, Dynein-dynactin-BicDR1 and Dynein-dynactin-HOOK3 produce stall forces of ~6.5 and ~5 pN, respectively. They also show increased velocities of ~1.25 µm/sec, compared to ~0.86 µm/sec displayed by BicD2. This increase in velocity and force production seems a direct consequence of recruitment of two motors. The increase in velocity has been attributed to the ability of BicDR1 and HOOK3 to arrange the HC tails of two dynein motors in parallel, and to keep them in a rigid orientation, which would reduce the sideways and backward stepping behavior of a single dynein, and promote a unidirectional, faster and consistent movement along microtubules.

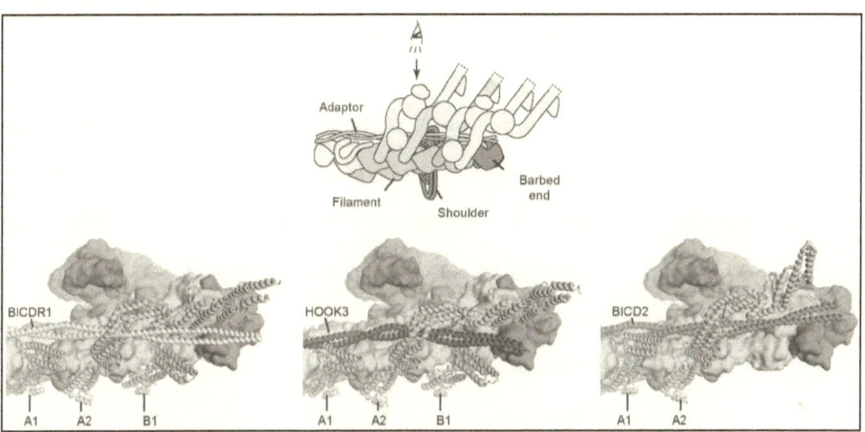

Figure 3. Some Dynein Adaptor Proteins Recruit Two Dynein Molecules for Added Force Generation and Processivity.

Comparative study of the molecular complexes of three different dynein adaptor proteins with dynein and dynactin show differential organization of the dynein-binding N-terminal coiled-coil domain in complex with the motor. HOOK3 and BicDR1 N-termini make a downward turn to contact the CapZαβ subunit of the dynactin complex, while the BicD2 N-terminus makes an upward turn to contact the Arp1A subunit of dynactin. This downward turn by the HOOK3 and BicDR1 N-termini allows for a second dynein recruitment, which in turn increases force production and processivity. Thus, dynein adaptor proteins show diverse mechanisms to recruit and control dynein motor activity. *(Adapted from Urnavicius et. al., 2018).*

The ability of BicDR1 and HOOK3 to interact with two dyneins has been attributed to their molecular contacts within dynactin. While the N-terminus of BicD2 makes an upward turn to interact with Arp1A, BicDR1 and HOOK3 N-termini make a downward turn to interact with CapZαβ. Only this latter mode of contact generates a contact site between dynactin and one of the tails of the second dynein molecule. Interestingly, while BicDR1 and HOOK3 both recruit two dyneins, they still generate different stall forces, suggesting that these adaptors have an added mechanism for modulating dynein behavior in cells.

Mechanism for Dynactin Activation of Dynein

It has been consistently observed that Dynein, on its own, exhibits poor binding to microtubules and weak processivity, which is drastically altered by its binding to dynactin. However, the molecular mechanism that regulates this switch in dynein efficacy had not been elucidated until recently. Recent work from the Carter lab shows that about 85% of dynein exists in an auto-inhibited form in mammalian cells (Zhang et al., 2017). This form is called the *Phi particle*. The dynein HC tail is highly flexible and is critical for maintaining the auto-inhibited form, and also for interacting with dynactin and cargo adaptors. This tail region consists of the N-terminal dimerization domain (NDD) followed by nine α-helical bundles. The ICs contact bundles 4 and 5, the LICs contact bundles 6 and 7, respectively. The N-terminal HC tail structure shows that the two HCs make three distinct contacts with each other, (i) The ~200 aa N-terminal NDD domain (between helical bundles 1 and 2), which can exist in either parallel (open, processive) or twisted (auto-inhibited) form, (ii) The Robl (*roadblock*) binding to IC N-terminus, and (iii) a novel contact close to LICs, between helical bundles 7 and 8.

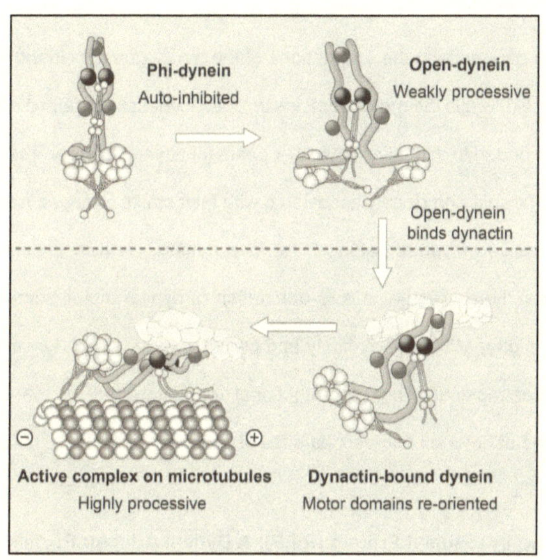

Figure 4. Mechanism of Activation of Dynein.
Cytoplasmic dynein exists in an auto inhibited 'phi particle' form in cells. Upon binding to dynactin, however, the dynein motor domains are oriented parallel to each other and the motor becomes processive *(Adapted from Zhang et. al., 2017)*.

Interestingly, the third contact close to LICs is only established when Dynein is in a Phi particle form. In the Phi particle state, dynein stalk shows a bulge in CC2 instead of forming a straight coiled-coil and the stalks essentially cross over one another. In this conformation, AAA1, the major ATP hydrolysis site, is tightly closed. Charge reversal mutations that disrupt the linker-AAA4 contact sites in Phi-dynein result in an increased (80%) population of 'open' dynein molecules, compared to wt dynein (~22%). These mutations also increase the mutant dynein's binding affinity to microtubules (the dwell times, however, remain unaltered). The mutant 'open' dynein also shows an increased association with dynactin and BicD2 N-terminus, suggesting that it is better primed to form a processive complex in this configuration. However, very importantly, the mutant 'open' dynein, by itself, does not exhibit long, processive movements, suggesting that even in an 'open' form, there is an added level of regulation to activate dynein, i.e. binding to dynactin. This is explained by the difference in the orientations of the motor domains in the 'open' *vs* dynactin-bound forms of dynein. While the motor domains in 'open' dynein face each other, they are parallel to each other when bound to dynactin. Thus, it appears that dynactin organizes the motor domains to make them less flexible and closely associated with their stalks arranged parallel to each other, facilitating a processive movement along MTs. In summary, we now understand that dynactin binding to an 'open' dynein results in a re-orientation of dynein motor domains and C-terminal stalks, to facilitate higher MT binding affinity and processivity. The Phi-dynein ensures that most of the dynein motor molecules are non-functional (or inactivated) in the cytoplasm and are selectively activated at different sub-cellular sites as needed.

1.8 Rab-interacting Lysosomal Protein (RILP): A Dynein Adaptor Protein

RILP is an α-helical extended coiled-coil domain containing dynein adaptor protein. RILP homologues have been identified in human, rat, mouse, chicken, *Caenorhabditis Elegans* and *Drosophila Melanogaster*, but not in yeast. While RILP is ubiquitously expressed in more than fifty different human tissues, its expression varies significantly, with high mRNA expression found in Heart, skeletal muscles and liver; and relatively low expression observed in brain and spinal cord tissue. I find RILP protein to be well-expressed in both embryonic and adult rat brain tissues. RILP gene transcription produces two isoforms: a long, 1.8 kb isoform, which is not expressed in brain tissue, and a short isoform, which is selectively expressed in brain tissue (Bucci, De Gregorio, & Bruni, 2001).

RILP has been previously established to recruit cytoplasmic dynein to Rab7-positive late endosomes (LEs) in non-neuronal cells (Cantalupo, Alifano, Roberti, Bruni, & Bucci, 2001; Ricardo Celestino et al., 2019; Jordens et al., 2001; Scherer et al., 2014).

FRAP (Fluorescence Recovery After Photobleaching) experiments have demonstrated that RILP recruitment to GTP-bound, membrane associated form of Rab7 results in retention of the GTP-bound Rab7 on vesicular membranes and impedes GTP hydrolysis on Rab7 (Jordens et al., 2001). This results in longer retention of Rab7-GTP on the vesicular membrane in presence of RILP. RILP has also been proposed to participate in extension of membrane protrusions from phagosomes, with pulling forces applied through RILP-associated dynein-dynactin motor complex along microtubules. These protrusions have been proposed to facilitate the fusion of the phagosomal compartment with lysosomes, suggesting an unusual RILP role in facilitating membrane fusion (Harrison, Bucci, Vieira, Schroer, & Grinstein, 2003).

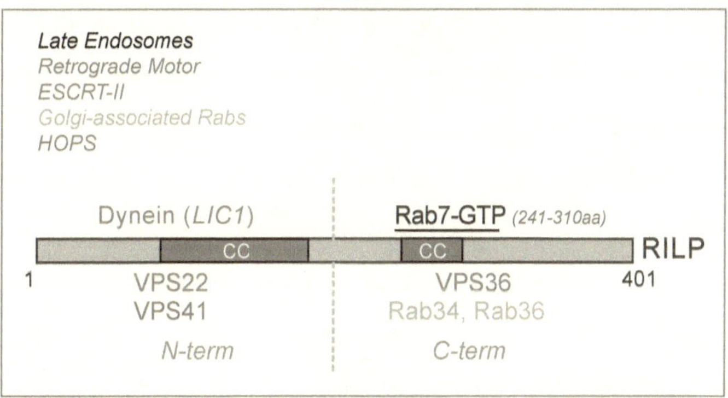

Figure 5. Previously known RILP Interacting Proteins and their Binding Sites.
Dynein directly binds to RILP N-terminus, while Rab7-GTPase binds the RILP C-terminus in a GTP-dependent manner. VPS22 and VPS36, two subunits of the ESCRT-II complex bind the N- and C-termini of RILP, respectively. VPS41, a subunit of the mammalian HOPS complex, binds to the RILP N-terminus. Finally, Golgi-associated Rab34 and Rab36 bind to the RILP C-terminus.

1.9 RILP Interactors

Several studies have identified a number of novel interactors for RILP in mammalian non-neuronal cells. Some of these interactions are briefly described below:

RILP interaction with Rab7-GTP

RILP was first identified as a Rab7-GTP binding protein. The Rab7-binding domain in RILP has been mapped to the C-terminal 241-320 amino acids. The crystal structure of RILP C-terminus and Rab7 complex shows a 2:2 symmetry, where a RILP homodimer binds two Rab7-GTPs simultaneously (Wu, Wang, Loh, Hong, & Song, 2005). GDP-bound Rab7 fails to bind RILP, suggesting that only the active, membrane-bound form of Rab7 is capable of recruiting RILP.

RILP C-terminal region spanning amino acids 273-333 has been shown to be necessary for RILP recruitment to lysosomal compartment. This region is absent in two other RILP family proteins, RILP-like 1 and 2 (T. Wang, Wong, & Hong, 2004), and thus, is considered necessary and sufficient for RILP recruitment to lysosomes, through Rab7 and possibly, Rab34 (Golgi-specific). RILP depletion has been reported to impede turnover of EGF-bound EGFR in non-neuronal cells. In this case, EGFR accumulates in early endosomes. Thus, RILP function may play an important role in maturation of the endo-lysosomal compartment (Progida et al., 2007).

RILP interaction with ESCRT-II Complex Subunits VPS22 and VPS36

ESCRT-II complex consists of a tetramer of two molecules of VPS25, and one molecule each of VPS22 and VPS36. Yeast two-hybrid analysis, mammalian localization and immunoprecipitation studies have shown RILP N-terminus (85-201aa) to directly interact with VPS22, a component of the ESCRT-II complex (Progida, Spinosa, De Luca, & Bucci, 2006). A further study in mammalian non-neuronal cultured cells showed that RILP interacts directly with VPS22 and VPS36 through N- and C-termini respectively; but fails to colocalize with VPS25 (T. Wang & Hong, 2006). These novel interactors of RILP suggest a role in multivesicular body (MVB) formation, however, how essential this function of RILP is for cellular survival, has not been further characterized.

RILP interaction with the HOPS complex

The mammalian HOPS complex consists of six subunits (namely, VPS11, VPS16, VPS18,

VPS33a, VPS39 and VPS41) and is one of the two known GEFs for mammalian Rab7 GTPase.

The only other known GEF for Rab7 is Mon1-CCZ1 (Yasuda et al., 2016). RILP has been shown

to interact with all HOPS complex subunits, except with VPS33a (van der Kant et al., 2013). RILP

interacts with these subunits through its N-terminal dynein binding domain. RILP has distinct

residues within the N-terminus that specifically interact with different HOPS subunits. RILP is able

to cluster LEs in mammalian cells, and these clusters show a potent recruitment of HOPS complex

proteins, in particular, VPS11, VPS18 and VPS41.

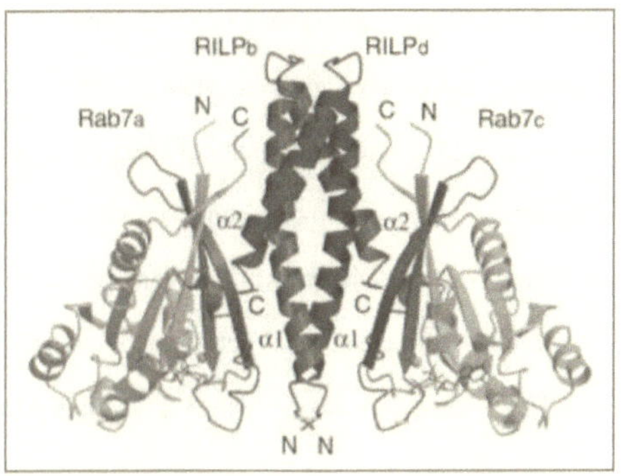

Figure 6. RILP Interaction with Rab7-GTP.
A crystal structure of the RILP C-terminal cargo binding domain interacting with Rab7-GTP shows
a 2:2 symmetry, where a dimer of RILP C-termini interacts with two molecules of Rab7-GTP. The
minimal Rab7-binding site in RILP has been mapped to amino acids 241-320.

It has been speculated that RILP may stabilize HOPS complex on the late endosomal membrane,

independent of Rab7. This has implications for how RILP might control the HOPS complex activity

to regulate the GTPase cycle of Rab7. Further work is needed to understand how RILP might

control endosomal maturation through its interactions with HOPS. It is possible that RILP recruits

HOPS complex to Rab7 on endosomal membrane and is required for maintaining Rab7 in a GTP-

bound form for a longer duration. A separate study also reported a direct *in vitro* interaction of

RILP N-terminal domain with VPS41, a HOPS complex subunit. This interaction was independent of Rab7 GTPase, further suggesting that RILP may recruit HOPS complex to regulate Rab7 GTPase activity on the endosomal membrane (Lin et al., 2014).

RILP interaction with the Vacuolar V-ATPase

Vacuolar-type H^+ ATPases are multi-subunit complexes that serve to regulate proton transport across the cellular membranes and thus, control the acidification of cellular compartments (Forgac, 2007; Sun-Wada, Wada, & Futai, 2004). RILP N-terminal domain interacts directly with the peripheral subunit V1G1 of the V-ATPase complex. While interacting with V1G1, RILP can also bind to Rab7, as well as p150[Glued] subunit of dynactin, and thus, may help stabilize the V-ATPase on Rab7-positive vesicles, while driving their retrograde transport (De Luca et al., 2014). Inhibition of RILP in non-neuronal cells results in accumulation of V1G1 in Giantin-positive Golgi compartments and a corresponding decrease in its localization to LAMP1-positive late endosomes. Thus, RILP activity might be critical for normal cellular localization of V1G1 to lysosomal compartments. Defects in RILP activity might impede lysosomal acidification and lysosomal functioning. The exact contribution of RILP to lysosomal function, however, needs to be further elucidated.

RILP interaction with Golgi-associated Rab proteins

RILP also interacts with two Golgi-associated Rab GTPases, Rab34 (T. Wang & Hong, 2005) and Rab36 (L. Chen, Hu, Yun, & Wang, 2010), respectively. Further studies have shown that Folliculin, a tumor suppressor protein implicated in Birt-Hogg-Dubé syndrome, *via* its C-terminus, directly binds RILP and, in turn, recruits it to Rab34-positive Golgi membranes that contact perinuclear lysosomes. Thus, Folliculin-RILP-Rab34 tripartite complex restricts lysosomal motility in the peri-nuclear region upon amino acid or serum starvation (Starling et al., 2016).

Multiple RILP Roles in Pathogen Clearance

From our lab's prior work (Scherer et al., 2014; Tan, Scherer, & Vallee, 2011) and that of others (Guignot et al., 2004; Harrison et al., 2004; Seto, Matsumoto, Tsujimura, & Koide, 2010; Starr, Ng, Wehrly, Knodler, & Celli, 2008; Sun et al., 2007; Wozniak, Long, Jones-Jamtgaard, &

Weinman, 2016), RILP has been implicated in endo-lysosomal mobilization during the host cell response to a variety of pathogens including adenovirus, Salmonella, Mycobacterium Bovis BCG and Tuberculosis, Brucella, Hepatitis C virus, as well as bacteria. The following section briefly reviews the variety of mechanisms adapted by different pathogens to either evade RILP-positive acidic compartments or to hijack these compartments to establish a safe, subcellular niche for survival and multiplication.

Adenovirus

Adenovirus infection activates Protein Kinase A (PKA) in host cells, which then phosphorylates dynein LIC1 (*light intermediate chain 1*) subunit. This phosphorylation of LIC1 prevents its binding to RILP. As a result, upon adenoviral infection, dynein gets dissociated from RILP-positive lysosomal membranes, and these vesicles slowly diffuse anterogradely towards cell periphery. This behavior of lysosomes may have an important cellular function in host cell defense against pathogen infection: The anterogradely moving lysosomes might intercept the invading virus, and target It for swift degradation (Scherer et al., 2014).

Hepatitis C Virus

HCV and Sendai viruses, upon infecting host cells, cleave intracellular RILP to generate a shorter version of RILP that lacks the N-terminal dynein motor binding domain (Wozniak et al., 2016). This C-terminal dynein binding deficient portion of RILP still associates with HCV and Sendai containing Rab7-positive vesicles but fails to recruit dynein for their retrograde transport to lysosomes. In absence of the retrograde motor, these vesicles show predominantly kinesin-driven anterograde motility, which prevents viral degradation by lysosomes and facilitates increased secretion of the virus from the host cell.

Salmonella

Salmonella Enterica secretes a type 3 effector SopD2, which hydrolyses GTP from the membrane bound Rab7-GTPase, but in doing so, immobilizes Rab7 onto the endosomal membrane (D'Costa et al., 2015). The Rab7-GTP hydrolysis prevents recruitment of RILP or

FYCO1 (kinesin adaptor) to salmonella containing vesicles (SCVs). In SopD2 expressing cells, RILP and FYCO1 remain mostly cytosolic and the microtubule-based motors are not recruited to the SCVs for their directional transport.

During the later stages of *Salmonella typhimurium* infection, this pathogen secretes SifA, a protein that competes with RILP to bind to Rab7-GTP. SifA prevents RILP association with endosomes, and the subsequent RILP-mediated dynein recruitment and retrograde transport of salmonella-containing compartments (Harrison et al., 2004). SifA secreting cells exhibit an anterograde transport of salmonella-containing vesicles, away from the degradative compartments.

Mycobacterium tuberculosis and Mycobacterium bovis

M. Tuberculosis survives and replicates within the phagosomal compartment of host macrophages. This pathogen preferentially recruits CD63 to phagosomal membranes while preventing RILP recruitment through a yet unknown mechanism (Seto et al., 2010). Absence of RILP on the *M. tuberculosis* containing phagosomes prevents their retrograde transport, and blocks their maturation through fusion with late endosomes and lysosomes. This behavior helps the pathogen escape the acidic compartment and survive longer within the host cell.

M. tuberculosis and *M. bovis* both reside within the phagosome in macrophages, and prevent its fusion with late endosomal and lysosomal compartments. This is effected by a Mycobacterium secreted factor that deactivates Rab7 to lock it in its GDP-bound form, which prevents Rab7 recruitment of RILP which would promote transport of phagosomes toward the acidic compartment and fusion with lysosomes (Sun et al., 2007).

Brucella

Brucella abortus, upon infecting the host cell, gets trafficked in a membrane-bound organelle called 'Brucella containing vacuole' (BCV), which rapidly acquires late endosomal markers such as Rab7, LAMP1 and RILP, and shows proteolytic activity for up to twelve hours post-infection (Starr et al., 2008). This maturation of BCV is required for its eventual passage to the ER, where it forms a replicative organelle derived from the ER membrane. The overexpression of RILP prevents BCV delivery to the ER, indicating that the BCV adopts RILP and other endosomal proteins for its survival and eventual replication within the host cell.

The examples discussed here show a wide variety of ways in which intracellular RILP activity is either inhibited or selectively adapted by the invading pathogens to promote their own survival and replication. Understanding the range of RILP functions, and the molecular details of RILP-Rab7 interaction may provide further insights into how RILP activity can be modulated to combat pathogen infections in mammalian cells.

RILP Roles in Cholesterol Transport

High cholesterol levels in cells induce formation of a multi-subunit complex containing ORP1L-Rab7-RILP-HOPS (van der Kant et al., 2013). In this condition, RILP is able to potently recruit the motor and transport LEs towards the lysosomal compartment (Despite the presented evidence, this raises questions as to the binding site availability of several HOPS subunits that RILP N-terminus has been shown to directly interact with, and the dynein motor, which also interacts within the RILP N-terminus). In low cholesterol conditions, ORP1L interacts, through its FFAT domain, with the ER VAP (Vamp-associated proteins) proteins. This is accompanied by the loss of interaction between HOPS, dynein and RILP. In this case, the retrograde transport by RILP is abrogated (Wijdeven et al., 2016).

As discussed in this section, RILP interacts with several mammalian proteins through distinct domains. However, the exact cellular roles played by RILP, and whether they are essential for cell survival and maintenance has remained poorly understood. In particular, RILP roles in the nervous system are completely unknown. RILP contributions to cellular signaling pathways are also not clear.

1.10 Novel Roles for RILP in Neuronal and Non-Neuronal Autophagy

We have discovered a novel autophagy mechanism which mediates neuronal autophagosome biogenesis and microtubule-based transport in neurons. We find that RILP controls these processes through direct sequential interactions with the core autophagy proteins ATG5 and LC3, as well as dynein. As discussed in the previous section, RILP has been previously established to recruit cytoplasmic dynein to Rab7-positive late endosomes (LEs) in non-neuronal cells (Cantalupo et al., 2001; Jordens et al., 2001; Scherer et al., 2014). From our own prior work (Scherer et al., 2014; Tan et al., 2011) and that of others (Harrison et al., 2004; Seto et al., 2010; Starr et al., 2008; Wozniak et al., 2016), RILP has also been implicated in endo-lysosomal mobilization during the host cell response to adenoviral and bacterial infection, which is reminiscent of xenophagy. Based on the existing evidence, we speculated that RILP might specifically contribute to the autophagy pathway, a possibility not so far explored.

We report that RILP expression and recruitment to autophagic membranes is specifically upregulated upon induction of autophagy through mTOR inhibition. We find that, RILP directly binds to the well-known isolation membrane protein ATG5, and facilitates the maturation of isolation membranes into fully-formed autophagosomes (APs), through a dynein-independent mechanism. We find that RILP then associates with the APs via LC3, a novel interaction mediated by three RILP C-terminal LIR (LC3-binding) motifs we identify in this study. Using RNAi and mutational analysis we find the LIR motifs to be both necessary and sufficient for RILP recruitment to autophagosomes, their retrograde transport, and the autophagic clearance of p62/Sequestosome-1. Together, our data lead to a model for RILP as a master regulator of mTOR-sensitive neuronal autophagy, controlling autophagosome biogenesis, transport and clearance.

CHAPTER 2. ROLES FOR DYNEIN ADAPTOR RILP IN NEURONAL AUTOPHAGY PATHWAY

Autophagosome biogenesis is initiated at PI3P-enriched nucleation sites within the ER and other membranous organelles (Axe et al., 2008; Derubeis, Young, Jia, Robey, & Fisher, 2000; Uemura et al., 2014). Isolation membranes expand from these sites to engulf ubiquitinated cytoplasmic cargoes. Closure of the isolation membranes yields double-membraned LC3-positive autophagosomes (APs) (Carlsson & Simonsen, 2015; Dooley et al., 2014; Mizushima, Yoshimori, & Ohsumi, 2011). In mammalian cells, the APs are transported retrogradely along microtubules, and fuse with Rab7-decorated late endosomes (LEs) and ultimately lysosomes, for cathepsin-mediated degradation and recycling of the autophagosomal contents (Bright, Davis, & Luzio, 2016; Gao, Langemeyer, Kummel, Reggiori, & Ungermann, 2018; Jia, Guardia, Pu, Chen, & Bonifacino, 2017). mTOR kinase serves a critical role in sensing nutrient deprivation and other forms of stress. Inhibition of mTOR activates autophagy and prolongs cell survival. However, how mTOR regulates neuronal autophagy remains a critical, but poorly understood question (Joungmok Kim et al., 2011; Maday & Holzbaur, 2016; Robert A. Saxton & Sabatini, 2017).

Cytoplasmic dynein (Dynein 1) is a microtubule minus end-directed motor protein responsible for most, if not all, interphase retrograde transport and numerous other aspects of cell physiology. How dynein is recruited to its remarkably broad range of cargoes is incompletely understood, as are mechanisms for regulating dynein behavior in response to diverse physiological changes. Autophagosomes (APs) are among the known dynein cargoes, but the full extent of dynein's contributions to autophagy remains poorly understood. Autophagy is critical for the cellular response to stress, in the form of starvation, injury, toxic deposits, infection, and other insults. My work has now identified an array of novel roles for the dynein adaptor Rab-interacting lysosomal protein RILP in nascent (prior to fusion with LEs) autophagosome formation, transport, and turnover. This behavior is mTOR-responsive and regulates turnover of Sequestosome-1

(SQSTM1/p62), which has been reported to aggregate in several neurodegenerative diseases (Caccamo, Ferreira, Branca, & Oddo, 2017b; Narendra, Kane, Hauser, Fearnley, & Youle, 2010). Together, our findings identify a novel, basic pathway for controlling autophagic progression, with broad implications for understanding cellular changes due to aging and disease.

2.1. Novel RILP Roles in Autophagy

We sought to find out the physiological roles for RILP in neuronal and non-neuronal cells. Given RILP's association with Rab7-positive compartments, and its involvement in several pathogen invasion pathways, we speculated that RILP might specifically contribute to the autophagy pathway, a possibility so far unexplored. We hypothesized that RILP might control the autophagy-endosomal axis to regulate turnover of cellular substrates as well as invading pathogens.

2.2. RILP Associates with Autophagosomes through C-terminal LC3-Interacting Regions

Although RILP has been implicated in several host defense mechanisms, a RILP role in autophagy has not been investigated. To address this issue, we tested for co-distribution and co-behavior of RILP with autophagosomal membranes in neuronal and non-neuronal cells. We expressed GFP-tagged RILP in HeLaM and C6 glioma cells and co-immunostained for endogenous Rab7 or LC3. Endogenous Rab7 showed clear colocalization with a subset of RILP vesicles. Interestingly, LC3 also showed striking colocalization with RILP-positive structures in both cell lines (Fig 8 k-m). To test whether RILP also associated with LC3-positive APs in neurons, we monitored the relative behavior of GFP-RILP in DIV (days-*in-vitro*) 8 rat embryonic cortical neurons expressing RFP-LC3, using high temporal resolution dual-channel live imaging. We observed extensive punctate colocalization of RILP with LC3-positive autophagosomes (APs) (88% of RILP vesicles were LC3-positive). The RILP-positive APs were highly motile, predominantly moving in the retrograde direction (49.14% RILP-positive APs) consistent with dynein-driven behavior (Fig 7).

To determine the molecular basis for RILP-LC3 colocalization, we first tested whether the Rab7-binding site in RILP was also responsible for its recruitment to autophagosomes. We introduced a single point mutation (Met305→Ala) within the Rab7-binding site, which has been previously reported to disrupt RILP-Rab7 binding (Wu et al., 2005). As expected, dual-color live imaging in DIV8 cortical neurons showed a striking decrease in localization of the Rab7-binding mutant RILP with mCh-Rab7-positive LEs (Fig 7c, e-g) (94% wild type *vs* 11.11% Rab7-mutant RILP), though this mutant remained fully able to localize to and transport RFP-LC3-positive APs (80.54% RILP-LIR mutant vesicles were Rab7-positive) (Fig 7d). Thus, the Rab7-binding site appears dispensable for RILP recruitment to APs. To further define the basis for RILP-AP binding, we examined the human RILP amino acid sequence for LC3- Interacting Regions (LIRs) (Alemu et al., 2012; Kalvari et al., 2014; Lamark, Kirkin, Dikic, & Johansen, 2009), which have been identified in a number of ATG family proteins, and are required for their interaction with LC3. We found one canonical and two F-type (Phenylalanine-containing) LIRs in close proximity to, but not overlapping with the Rab7-GTP binding site within the C-terminal "cargo binding" domain of RILP (Fig 7a, b). To test the physiological significance of these motifs, we introduced missense mutations in the first aromatic amino acid residue to change each to valine. To determine the effects of the mutations in rat cortical neurons, we co-expressed single, double or triple LIR-mutant versions of GFP-tagged RILP, along with RFP-LC3. Relative to *wt* RILP, we observed a graded loss in RILP LIR mutant – AP colocalization (88.4% *wt* RILP, 67.83% LIR1- mutant, 61.98% LIR2- mutant, 32.3% LIR3- mutant RILP vesicles colocalized with LC3), which was dependent on increase in the number of LIR mutations in RILP (Fig 8 a-d). These results showed that all three LIR motifs were functional and, indeed, necessary for RILP recruitment to APs. The triple LIR-mutant RILP showed the most potent loss in AP localization, so for subsequent analysis we predominantly used this construct, here on referred to as "LIR3-mutant RILP". This construct also caused a marked decrease in the frequency of retrograde AP transport (49.14% wt *vs* 29.3% LIR3- RILP vesicles retrogradely transported) (Fig 7g). Using biochemical analysis, we also observed an ~50% reduction in LC3-binding by the LIR3-mutant relative to *wt* RILP as judged by co-immunoprecipitation (Fig 8 e, f). Despite the decrease in localization to APs, the LIR3-mutant RILP still colocalized and comigrated with Rab7-positive late endosomes (LEs) as efficiently as *wt* RILP (94% wt *vs* 80.5% LIR3- mutant RILP colocalized with Rab7) (Fig 7 k). The velocities and travel distances of residual APs and LEs that still exhibited retrograde motility remained unaltered relative to *wt*, suggesting normal dynein behavior for these vesicles (Fig 8 g-h). These results show that RILP recruitment to APs occurs through discrete C-terminal LIR motifs, independent of the Rab7-binding site. Furthermore, the LIRs are necessary and sufficient for RILP recruitment to APs, independent of whether Rab7-binding is impaired. Together our results implicate RILP in dynein recruitment to both early (pre-LE fusion) as well as late (post-LE fusion) autophagosomal intermediates *via* RILP's distinct LC3- and Rab7-binding sites.

Intriguingly, expression of the LIR3-mutant RILP resulted in a striking decrease in the total number of APs per unit length of axon compared to *wt* RILP (0.047 *vs* 0.009 APs/ unit length in WT vs LIR3- mutant RILP) (Fig 7h). In these neurons, RFP-LC3 exhibited a predominantly diffuse cytoplasmic distribution. The Rab7-binding mutant RILP, in contrast, had no effect on the abundance of LEs or APs in neurons (0.045 APs/ unit length in WT vs Rab7 mutant RILP) (Fig 7n). These results suggest that, in addition to playing a key role in the retrograde transport of autophagosomes in neurons, the RILP-LC3 interaction might be involved in their biogenesis as well.

NT
CT
A
Dynein, Dynactin
Rab7-GTP
CC
CC
RILP
1
LC3
LC3 LC3
401
B
246-251
EAGQCRFSREEFEQI
320-325
EEAESSEDEAGPWILL
341-346
PPPPESKIQSFFGL
Retro
Antero
Retro
Antero
Retro
Antero
C
RILP
D
RILP LIR3- Mutant
E
RILP Rab7 Mutant
LC3
LC3
LC3
RILP
RILP LIR3- Mutant
RILP Rab7 Mutant
LC3
LC3
LC3
F
Vesicles (%)
120
100
80
60
40
20
0
**
***
LC3 -
LC3 +
RILP
WT
LIR3-
Rab7-
G
LC3 Vesicles (%)
80
60
40
20
0
RILP WT
LIR3-mut RILP
Rab7-mut RILP
**
*
Retro
Antero
Stat
Bidir
H
LC3 Vesicles / µm Axon
0.06
0.04
0.02
0.00
***
RILP
WT
LIR3-
Rab7-
Retro
Antero
Retro
Antero
Retro
Antero
I
RILP
J
RILP Rab7 mutant
K
RILP LIR3- mutant
Rab7
Rab7
Rab7
RILP
RILP Rab7 mutant
RILP LIR3- mutant
Rab7
Rab7
Rab7
L
Vesicles (%)
120
100
80
60
40
20
0
***
***
Rab7 -
Rab7 +
RILP
WT
LIR3-
Rab7-
M
Rab7 Vesicles (%)
80
60
40
20
0
RILP WT
LIR3- mut RILP
Rab7-mut RILP
*
Retro
Antero
Stat
Bidir
N
Rab7 vesicles /µm Axon
0.10
0.08
0.06
0.04
0.02
0.00
RILP
WT
LIR3-
Rab7-

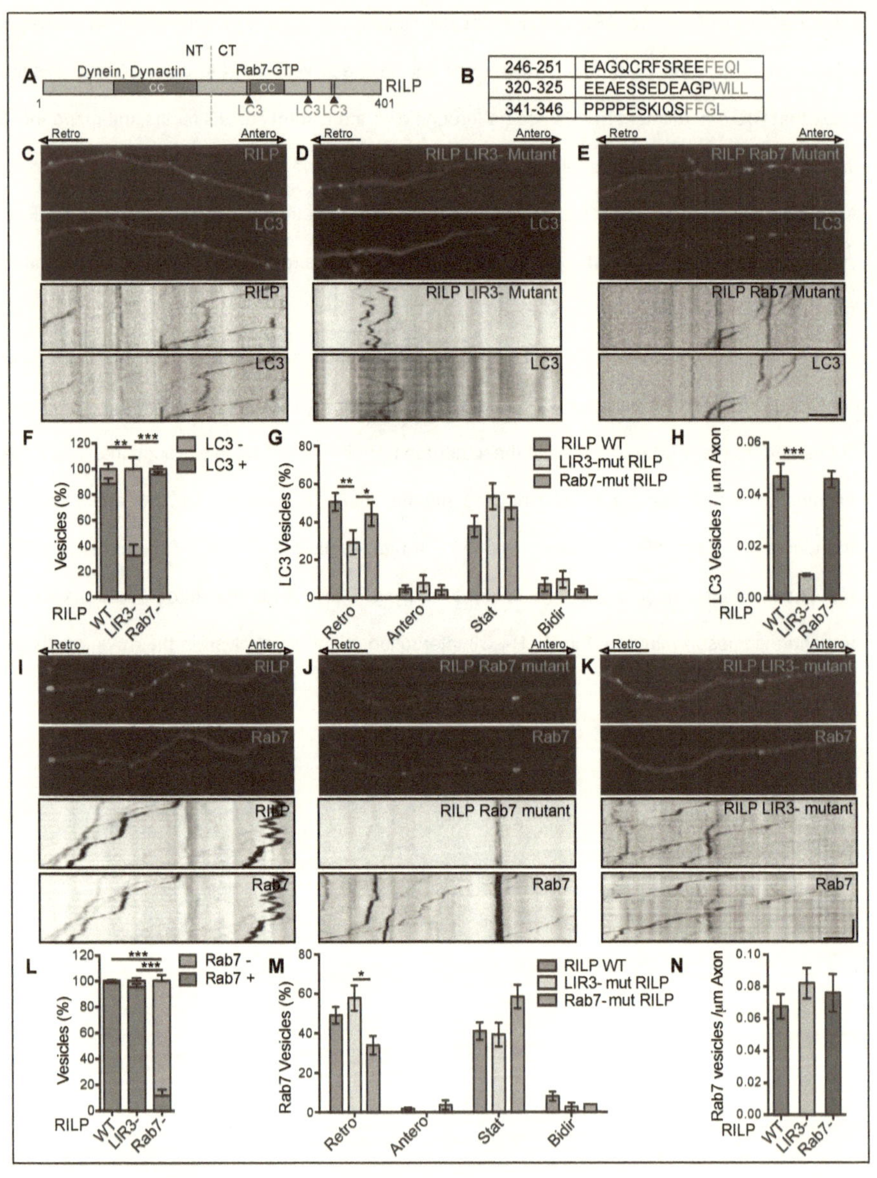

48

Figure 7. RILP Regulates Retrograde Axonal Transport of LC3-positive Autophagosomes in Neurons.

(A) RILP domain organization and interactions. RILP is an elongated α-helical coiled- coil (CC) –containing protein known to interact with cytoplasmic dynein and its regulator dynactin, as well as with the late endosomal marker Rab7. This study also shows RILP to interact with the autophagosomal membrane protein LC3. Three C-terminal LC3-interacting regions (LIRs) identified in this study are indicated. **(B)** LIR amino acid sequences, the second of which is a canonical (WxxL) motif, whereas the others are F-type (phenylalanine-containing). **(C-E)** GFP-RILP and RFP-LC3 were co-expressed in DIV (Day *in vitro*) 7-8 rat cortical neurons and monitored by dual-channel live time-lapse imaging. Single frames showing individual axons are presented at top for **(C)** *wt*; **(D)** LIR3- mutant; and **(E)** Rab7-binding mutant RILP *vs.* RFP-LC3. Shown below are kymographs indicating the change in AP position within the axon with time. **(F-H)** Quantification of AP behavior. **(C)** *wt* RILP colocalized and co-migrated with RFP-LC3-positive APs (n=31), predominantly in the retrograde direction. **(D)** In contrast, the LIR3- mutant RILP showed reduced localization and retrograde transport, and AP number was also markedly decreased (n=40). **(E)** Rab7-binding mutant RILP colocalized and co-migrated with APs (n=26). **(F)** Relative colocalization of *wt* and mutant RILP with APs. Rab7-binding site is dispensable for AP binding. **(G)** Decreased retrograde AP transport observed with LIR3- mutant RILP, but not with *wt* and Rab7-binding mutant RILP. **(H)** Decreased number of APs per unit length of axon in LIR3- mutant RILP expressing neurons suggests a role in AP formation. **(I-K)** GFP-RILP and mCh-Rab7 expressing DIV7-8 rat cortical neurons were monitored by live time-lapse imaging. Single frames showing individual axons are presented for **(I)** *wt*; **(J)** Rab7-binding mutant; and **(K)** LIR3-mutant RILP *vs.* mCh-Rab7. Shown below are kymographs and **(L-N)** quantification of LE behavior. **(I)** *wt* RILP colocalized and co-migrated with mCh-Rab7-positive LEs (n=31), predominantly in the retrograde direction. (n=30). **(J)** In contrast, the Rab7-binding mutant RILP showed reduced localization to LEs (n=16). **(K)** LIR3-mutant RILP colocalized and co-migrated with LEs (n=20). **(L)** Relative colocalization of *wt* and mutant RILP with LEs. LIRs are dispensable for Rab7-binding. **(M)** Decreased retrograde LE transport observed with Rab7-binding mutant RILP, but not *wt* and LIR3- mutant RILP. **(N)** No effect of Rab7-binding mutant RILP on LE number in axons. scale bars: x = 5 μm; y = 30 sec. P-values: *=0.05, **=0.01, ***=0.001.

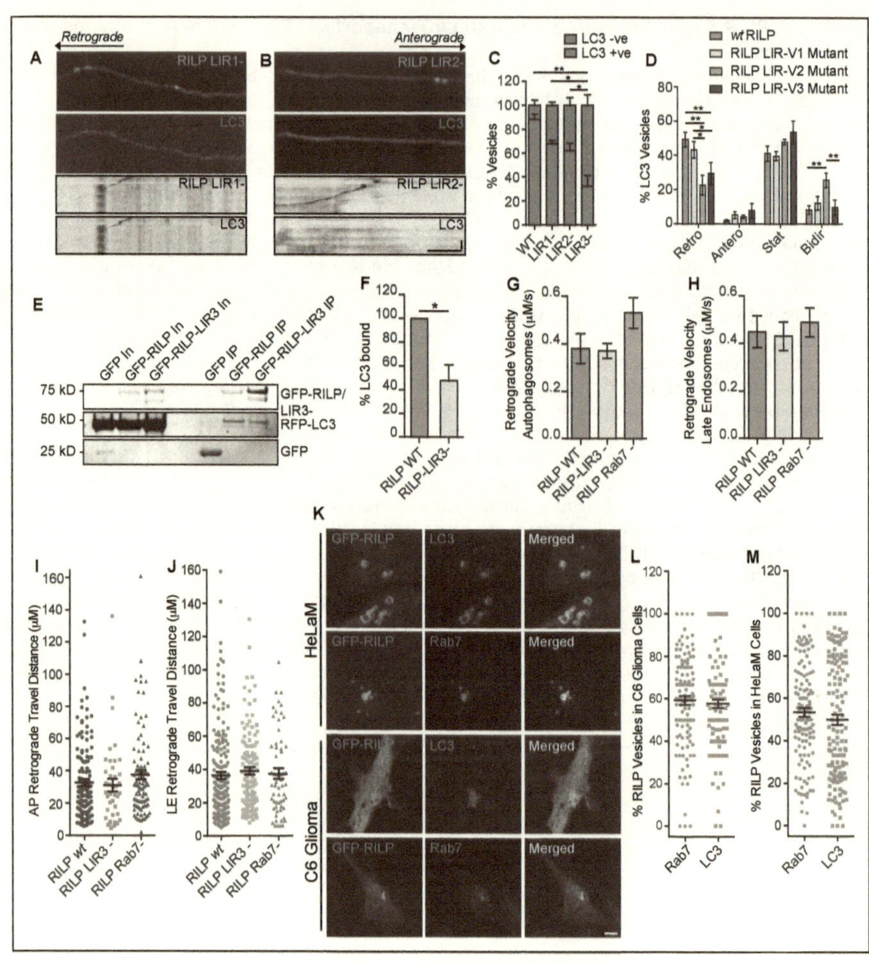

Figure 8. Effects of RILP LIR Motif Mutations on AP Binding and Motility.
In order to determine the contribution of individual LIR motifs in RILP recruitment to autophagosomes, we point mutated either one (WILL), two (WILL and FEQI) or all three (WILL, FEQI and FFGL) LIR motifs in RILP. These single, double or triple LIR-mutant constructs are referred to as RILP LIR1-mutant, LIR2-mutant and LIR3-mutant, respectively. LIR mutant GFP-RILP was co-expressed with RFP-LC3 in DIV7-8 rat cortical neurons and monitored by dual-channel live time-lapse imaging. Single frames showing individual axons are presented at top for **(A)** RILP LIR1-mutant and **(B)** RILP LIR2- mutant *vs.* RFP-LC3. Shown below are kymographs indicating the change in AP position within the axon with time (scale bars: x = 5 μm; y = 30 sec). **(C)** Relative localization of RILP *wt* and LIR mutants with RFP-LC3 (Data for *wt* RILP and LIR3-mutant are reproduced here from Fig. 1(F) for comparison). **(D)** LIR1- and LIR2- mutant RILP show reduced retrograde transport of APs and an increase in stationary APs along the axon. **(E-F)** *wt* or LIR3-mutant GFP-tagged RILP and RFP-LC3 were co-expressed in rat C6 glioma cells and the levels of LC3 co-immunoprecipitated were compared. LIR3-mutant RILP immunoprecipitated ~50% less RFP-LC3 than *wt* RILP. GFP empty vector was used as a negative control, and did not co-precipitate detectable levels of LC3 (n=3). **(G, H)** Retrograde velocities of *wt*, LIR3- mutant and Rab7-mutant RILP decorated **(G)** APs and **(H)** LEs are comparable, suggesting that the biophysical properties of the motor itself are unaffected upon RILP manipulation. **(I, J)** The average retrograde travel distances of LIR3-mutant and Rab7-mutant RILP are not substantially different than *wt* RILP, indicating that the dynein supercomplexes containing mutant RILP are functional, and the phenotypes observed are specifically due to loss of RILP-mediated recruitment of dynein to **(I)** APs and **(J)** LEs. **(K)** HeLaM and C6 glioma cells expressing GFP-RILP were immunostained for endogenous Rab7 and LC3, quantified in **(L)** and **(M)** for C6 glioma and HeLaM cells, respectively. P-Value: *=0.05.

2.3. RILP Depletion and LIR Mutations Inhibit Retrograde Autophagosome Transport, and Prevent Autophagic Clearance.

To test whether RILP plays an essential role in neuronal autophagy, we performed RNAi to deplete endogenous RILP in cortical neurons (Fig 9 a, b; Fig 10 g). This resulted in a marked decrease in the number of retrogradely transported APs (48.7% control *vs* 19.7% RILP KD) and an increase in stationary APs along the axon (39% control *vs* 75.3% RILP KD) (Fig 9 c-d, Fig 10 d-e). There was also a modest but detectable decrease in anterogradely transported APs, which might reflect functional interactions between retrograde and anterograde motor proteins (Ally, Larson, Barlan, Rice, & Gelfand, 2009; Yi et al., 2011). In agreement with our observations with the LIR3-mutant RILP, RILP RNAi also reduced axonal AP number (0.09 APs/ unit length in control *vs* 0.045 in RILP KD) (Fig 9 g), further supporting an unexpected and completely novel role for RILP in AP biogenesis. In keeping with the previously reported RILP-Rab7 interaction (Wu et al., 2005), RILP RNAi also resulted in a decreased retrograde axonal transport of Rab7-positive late endosomes in cortical neurons (39% control *vs* 10.1% RILP KD) (Fig 9 e-f, Fig 10 a-b). Finally, distances travelled by residual APs and LEs still exhibiting retrograde transport after RILP depletion appeared unaltered (Fig 10 c, f), confirming that dynein activity in these neurons remained normal.

To examine whether the RILP role that we identified in neuronal autophagy has a functional consequence in clearance of autophagic cargoes, we monitored the behavior of myc-tagged Sequestosome 1 (SQSTM1/ p62) (Zaffagnini et al., 2018a, 2018b) in RILP-depleted C6 glioma cells. We observed few myc-p62 puncta in the cytoplasm of control cells, but a striking increase in the number of p62 aggregates following RILP knockdown (3 in GFP *vs* 17 in RILP KD) (Fig 11 a-b, e). These results provide evidence for a functional RILP role in turnover of autophagic substrates. We also performed this assay in C6 glioma cells transiently expressing LIR3-mutant RILP vs. *wt* RILP or GFP alone.

We found no detectable difference in the number of p62 aggregates in GFP vs. *wt* RILP expressing cells; however, there was a substantial increase in the number of p62 aggregates in LIR3-mutant RILP expressing cells (Fig 11 c, d). We found further that the p62 aggregates were enriched for LC3 (56.1% aggregates) and Ubiquitin (57.2% aggregates) (Fig 11 f-g, j-k), but not for Rab7 (7.64% aggregates) or the aggresome marker, Vimentin (Fig 11 h-i, l-m) (Johnston, Ward, & Kopito, 1998; Moriya et al., 2015). Thus, in the absence of associated RILP, ubiquitinated substrates positive for p62 tend to accumulate in the cytoplasm and fail to reach the late endosomal compartment, further highlighting the role of RILP in the regulation of early stages in the autophagy pathway. Together, these results show that RILP is essential for autophagic turnover of p62/ Sequestosome-1, and this function requires the specific interaction of RILP with autophagosomes through its LIR motifs.

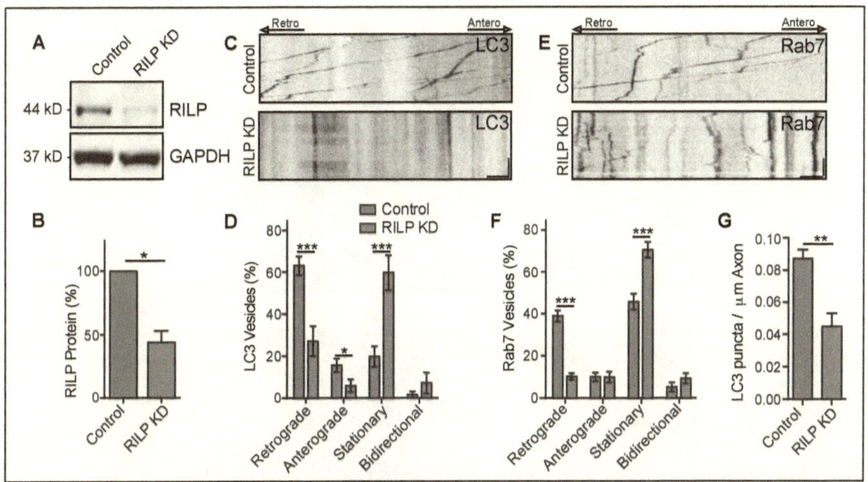

Figure 9. RILP is Essential for Axonal Autophagosome Transport.
RILP was knocked down in DIV3 rat cortical neurons using RNAi and its effects on AP and LE behavior were monitored by time-lapse imaging of **(C)** RFP-LC3-positive APs, or **(E)** mCh-Rab7-positive LEs (n >20). Kymographs indicate the change in axonal AP and LE position over time in control (top) and RILP RNAi (bottom) neurons (scale bars: x = 5 µm; y = 30 sec). **(A)** Western blot showing depletion of RILP protein upon 72 hr RILP RNAi in rat C6 glioma cells, quantified in **(B)**. Kymographs show processive retrograde transport of APs **(C)** and LEs **(E)** in control axons, which is inhibited upon RILP RNAi. **(D)** Quantitative analysis reveals specific RILP RNAi-mediated inhibition of retrograde AP and **(F)** LE transport. **(G)** Decreased number of APs upon RILP RNAi indicates a RILP role in AP formation. P-values: *=0.05, **=0.01 and ***=0.001.

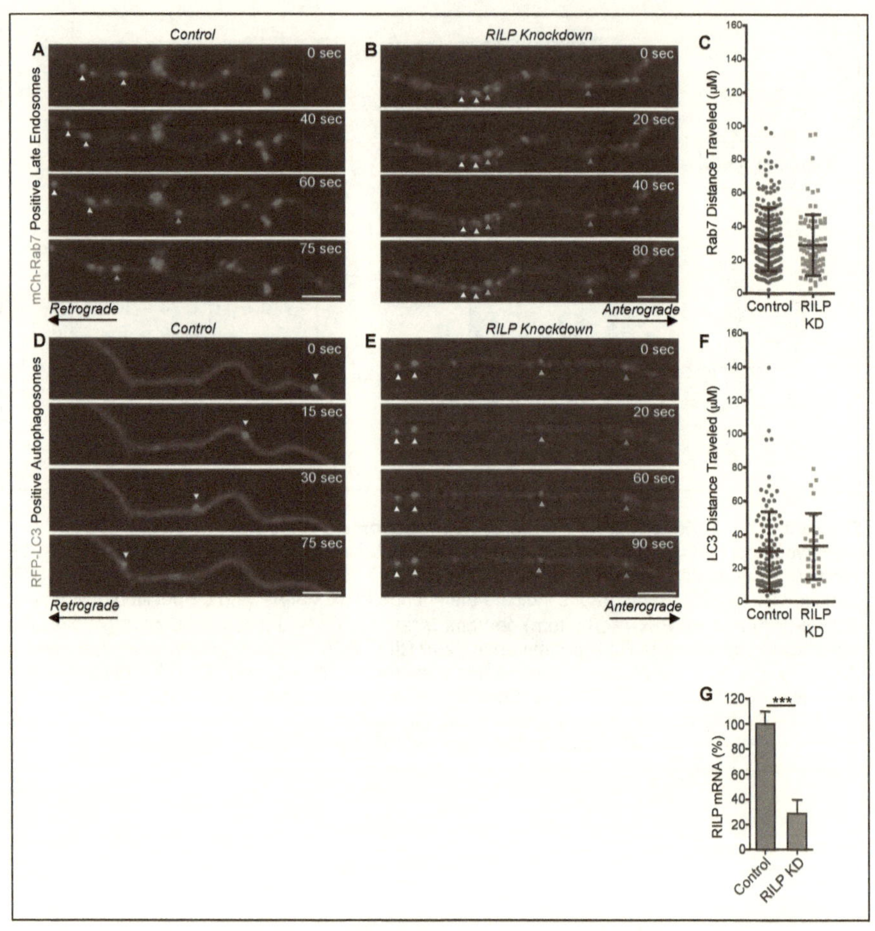

Figure 10. Effect of RILP RNAi on AP/LE Motility and Distance Travelled Along Axons.
RILP shRNA was co-expressed with either RFP-LC3 or mCh-Rab7 in DIV3 rat cortical neurons. Time-lapse series depicting late endosomal **(A-B)** and autophagosomal **(D-E)** behavior in control *vs* RILP RNAi cortical neurons. In control axons, mCh-Rab7 labeled LEs **(A)** and RFP-LC3 labeled APs **(D)** show processive retrograde transport. Individual puncta are labeled by arrowheads. Upon RILP RNAi, a substantial number of LEs **(B)** and APs **(E)** are stalled along the length of the axon (Scale Bars: 5µm). Distance travelled by **(C)** Late endosomes or Autophagosomes **(F)** in control *vs* RILP RNAi axons is not substantially different, although the total number of motile puncta in RILP RNAi neurons are substantially lower than control axons (quantified in Figure 2). **(G)** qRT-PCR of RILP mRNA levels following shRNA-mediated RILP knockdown in rat C6 glioma cells shows ~70% reduction. P value: ***=0.001.

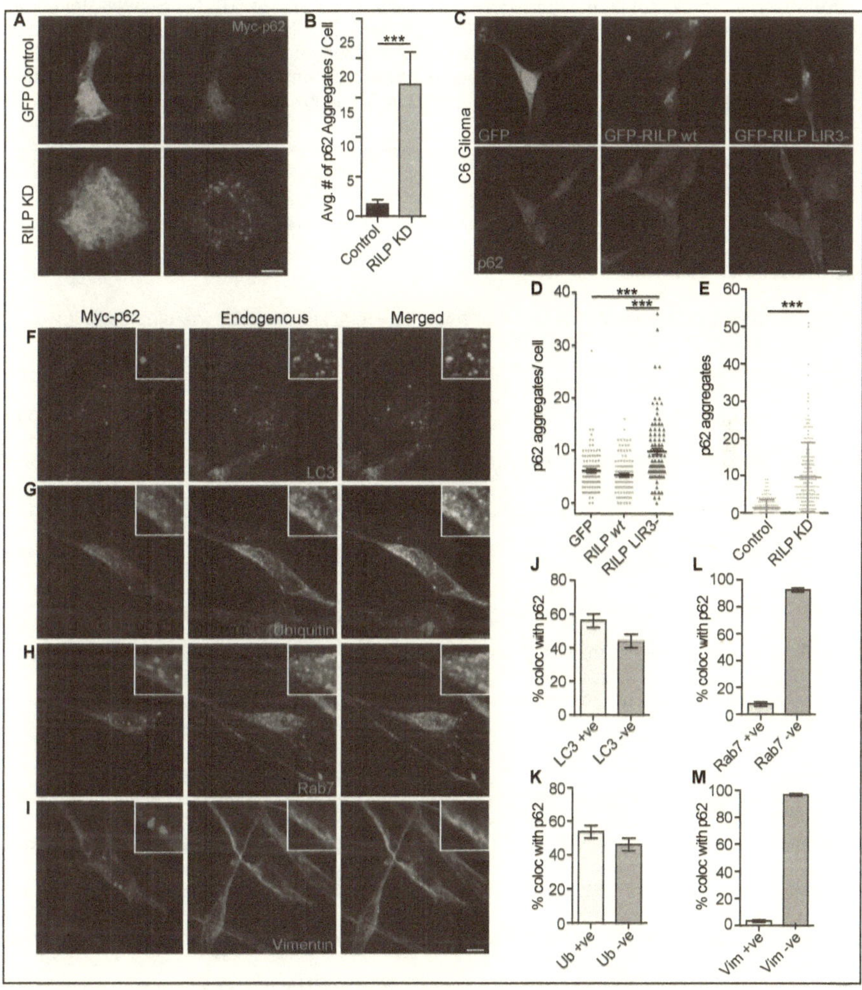

Figure 11. RILP Depletion Prevents Autophagic Clearance of p62/Sequestosome-1.
As a test for a physiological RILP role in autophagic turnover, we evaluated the distribution of Myc-tagged p62/sequestosome-1 (SQSTM1) with or without RILP RNAi in C6 cells. **(A)** p62 exhibited diffuse cytoplasmic distribution (top) in control cells. Upon RILP RNAi, p62 showed extensive aggregation (bottom), quantified in **(B)** (n>90). **(C, D)** GFP alone, *wt* RILP or LIR3-mutant RILP were transfected into C6 cells and endogenous p62 puncta were counted after 24 hours of expression (n=105 per condition). **(C)** Using immunostaining, comparable numbers of p62 puncta were detected in GFP and *wt* RILP expressing cells, but substantially higher numbers

were observed in LIR3-mutant RILP expressing cells, quantified in **(D)**. **(E)** Scatter plot shows relative increase in p62 aggregates upon RILP RNAi in C6 cells. **(F-M)** Further characterization of p62 aggregates shows that they are positive for endogenous LC3 **(F)** and ubiquitin **(G)**, but mostly negative for Rab7 **(H)** and aggresome marker vimentin **(I)**. Quantification of relative localization of p62 aggregates with LC3 **(J)**, Ubiquitin **(K)**, Rab7 **(L)** and Vimentin **(M)**. P-value: ***=0.001.

2.4. Induction of Autophagy by mTOR Inhibition Upregulates RILP Expression and Alters its Subcellular Behavior.

In view of the substantial range of roles we identified for RILP in autophagy, it was of interest to determine how RILP's behavior in cells was affected by mTOR function. For this purpose, we used a small molecule-mediated inhibition of mTOR kinase to induce autophagy. mTOR is the major nutrient level and stress sensor in cells, and is the key negative regulator of mammalian autophagy. Inhibition of mTOR by Rapamycin or the more potent inhibitor Torin1, is known to induce autophagy in non-neuronal cells, as monitored by formation of LC3-II positive autophagosomes and inhibition of S6K phosphorylation (Thoreen & Sabatini, 2009). Although mTOR kinase plays critical roles in processes underlying neurodevelopment and degeneration, the mechanisms by which it regulates neuronal autophagy remain unclear (Maday & Holzbaur, 2016).

To evaluate the effects of mTOR on RILP function, we treated C6 glioma cells and rat cortical neurons with Torin1. Inhibition of mTOR kinase activity and induction of autophagy were confirmed by a decrease in the level of S6K phosphorylation and an increase in the ratio of converted (lipidated) to full length LC3, respectively (Fig 12 a). We found that short-term (500nm, 6 hr) Torin1 treatment in C6 Glioma cells increased RILP mRNA levels by 2.36 +/- 0.45 -fold (Fig 12 b). There was also an upregulation in RILP protein levels as early as 2 hr post-Torin1 treatment (Fig 12 c). To determine whether mTOR-mediated upregulation was unique to RILP, we immunoblotted for other well-known motor protein adaptors, including the dynein adaptors HOOK1, HOOK3 and BicD2, as well as a kinesin adaptor FYCO1, the latter shown to bind LC3 and to recruit Kinesin-1 to APs (Pankiv et al., 2010; Raiborg et al., 2015) (Fig 12 d-g). However,

we detected no alteration in the total protein levels of these adaptors, which suggested that RILP

expression is specifically responsive to changes in mTOR activity in cells.

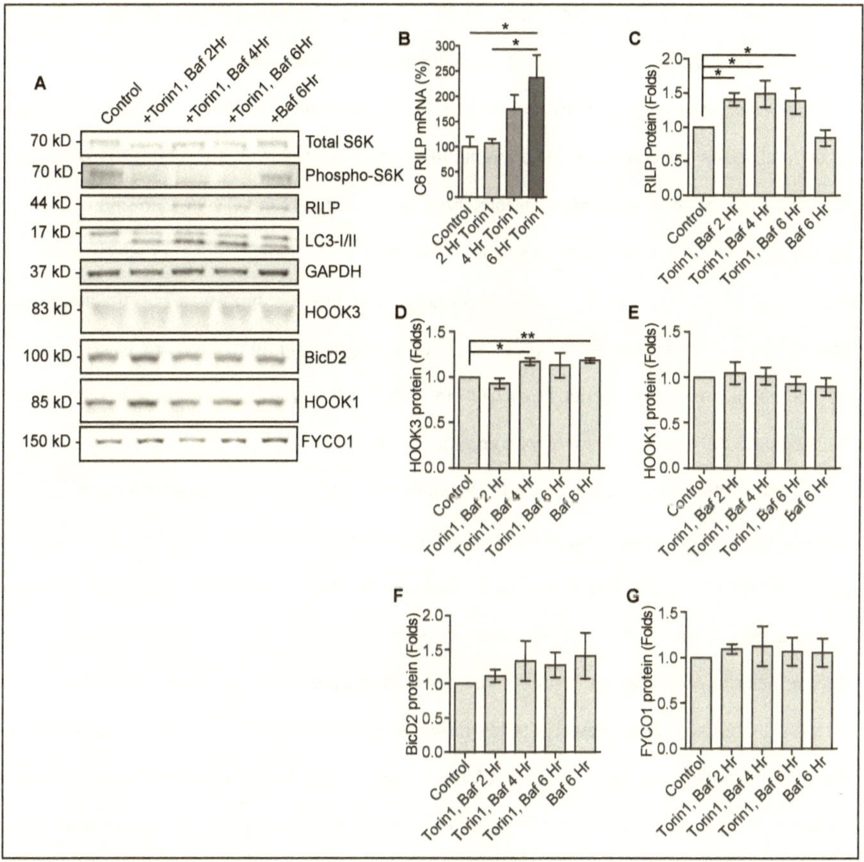

Figure 12. Effects of mTOR Inhibition on RILP Expression.
Rat C6 glioma cells were treated with Torin1 to induce autophagy, and effects on RILP and other
motor adaptors were determined using immunoblotting. **(A)** Western blot showing a decrease in
S6K phosphorylation, an increase in LC3-II/I ratio and levels of several adaptor proteins upon
induction of autophagy in C6 cells. **(B)** qRT-PCR analysis reveals a Torin1-induced increase in
RILP mRNA in glioma cells. **(C)** RILP protein levels increase as soon as 2 hours post-Torin1 plus
Bafilomycin A1 treatment, but not upon Bafilomycin treatment alone. No detectable changes in

levels of other dynein adaptors such as HOOK3 **(D)**, HOOK1 **(E)**, BicD2 **(F)** and kinesin adaptor FYCO1 **(G)** were observed. p-values: *$\leq$0.05, **$\leq$0.01

Next, we tested the effect of mTOR inhibition on endogenous RILP behavior using immunocytochemistry in rat NRK-F cells. In these cells, endogenous RILP and LC3 exhibited a mostly diffuse cytoplasmic staining, with occasional puncta observable, consistent with low levels of basal autophagy (Fig 13 c (i)). Short-term Torin1 treatment combined with Bafilomycin A1, which neutralizes lysosomal pH and prevents AP-LE fusion (Mauvezin & Neufeld, 2015), resulted in robust stimulation of autophagy, as observed by an increase in LC3-II/I ratio and a dramatic redistribution of RILP to discrete perinuclear puncta, a predominant subset of which were also positive for LC3 (Fig 13 (a-b), c (iv)). Importantly, this phenomenon was not observed upon Bafilomycin A1 treatment alone (Fig. 13 c (iii)), suggesting that RILP localization to AP membranes is specifically induced by mTOR inhibition, rather than reflecting a defect in fusion of RILP-positive APs with the lysosomal compartment (Fig 13 c (ii)).

We further tested whether RILP shows similar behavior in neurons as well. In cortical neurons treated with Torin1, we observed a more modest increase in RILP mRNA levels (Fig 13 d). We reasoned that this difference might reflect a higher basal level of mTOR activity in neurons due to a significantly higher concentration of insulin in the B-27 supplement added to neuronal culture medium compared to non-neuronal cell culture medium, which, itself stimulates mTOR (Brewer & Cotman, 1989; Brewer, Torricelli, Evege, & Price, 1993; Y. Chen et al., 2008; Vander Haar, Lee, Bandhakavi, Griffin, & Kim, 2007). Indeed, short-term Insulin withdrawal revealed a similar ~2.1 -fold Torin1-mediated increase in neuronal RILP mRNA levels, comparable to the effect we observed in glioma cells (Fig 13 d). We also found that cytoplasmic RILP behavior in these neurons was similar to that seen in NRK-F cells. In axons of untreated neurons, RILP exhibited a mostly diffuse cytoplasmic distribution, however mTOR inhibition resulted in a clear increase in RILP-positive vesicles, in some of which a clear lumen could be discerned (Fig 13 e).

Together, these results show that mTOR-dependent induction of autophagy increases RILP

expression and promotes its recruitment to newly forming autophagosomes.

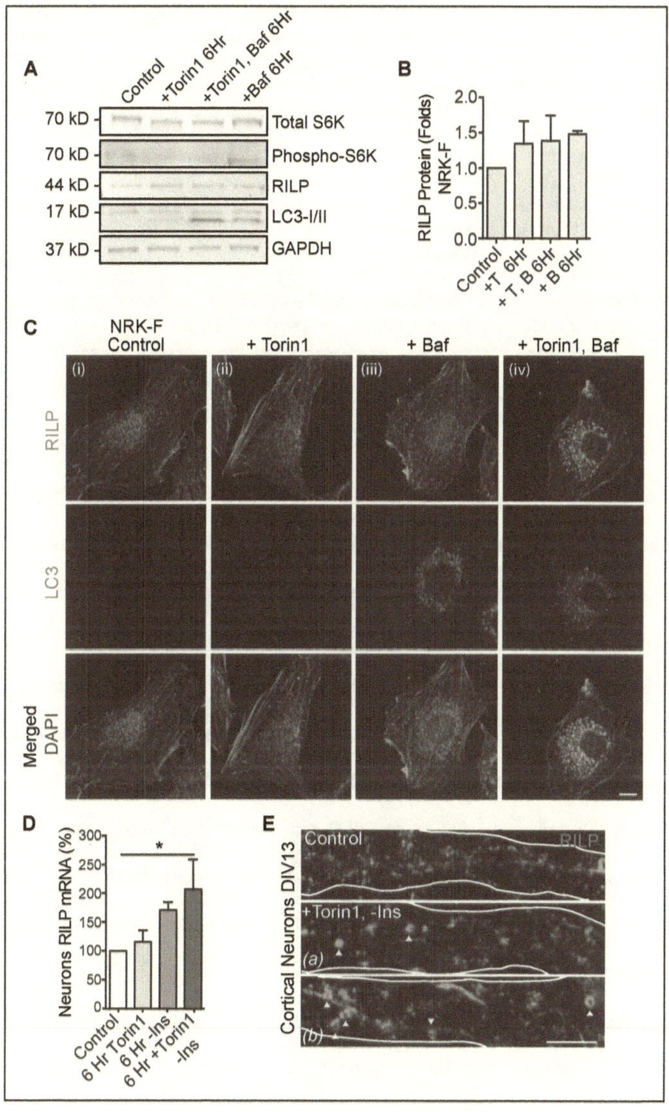

Figure 13. Effects of mTOR Inhibition on RILP Subcellular Behavior.
Rat NRK-F cells were treated with Torin1 (500nm) and Bafilomycin A1 (100nm) for a total duration of 6 hours. **(A)** Induction of autophagy was confirmed by decreased levels of S6K phosphorylation and an increase in the ratio of converted (lipidated) *vs* full-length LC3. **(B)** There is a modest but not significant increase in RILP protein levels in NRK-F cells. **(C)** Immunofluorescence analysis in control cells **(i)** shows mostly diffuse cytoplasmic distribution of endogenous RILP. **(ii)** Torin1 treatment alone did not show accumulation of RILP and LC3 dual-positive vesicles, which reflects a constant turnover of APs in the absence of Bafilomycin treatment as expected. **(iii)** Bafilomycin treatment alone resulted in an accumulation of LC3-positive APs but did not alter RILP cytoplasmic distribution. **(iv)** A brief Torin1 plus Bafilomycin treatment resulted in a dramatic redistribution of RILP onto vesicular structures, a subset of which were also positive for LC3 (n>150 cells for each condition). **(D)** DIV12-14 rat cortical neurons were treated with Torin1 to induce autophagy, and effects on RILP mRNA were determined. qRT-PCR analysis of RILP mRNA levels in DIV13 cortical neurons revealed a small increase from Torin1 alone, though 1.7- and 2.07- fold increases are observed upon insulin withdrawal with or without Torin1 treatment. **(E)** Endogenous RILP in DIV13 cortical neurons shows mostly diffuse cytoplasmic distribution, but Torin1-treatment combined with insulin withdrawal results in RILP recruitment to vesicular structures, some with a discernable lumen. Scale bars: 5 μm. p-value: *=0.5.

2.5. RILP is Recruited to Isolation Membranes through ATG5

In view of its unexpected effect on regulating autophagosome number, we hypothesized that RILP might be involved in the early stages of autophagy, i.e. in some aspect of biogenesis. To test this hypothesis, we used live imaging to probe for a RILP association with early, pre-closure autophagosomal intermediates in rat cortical neurons, using double-labeling for the nucleation site and isolation membrane markers DFCP1 and ATG5, respectively (Fig 14, Fig 15 a-e, j-k). RILP was present at each of these structures. Close inspection of the distribution pattern for RILP *vs* DFCP1 staining revealed RILP-positive membranes to be juxtaposed to and extending from the DFCP1 puncta in DIV7-8 neurons (Fig 15 j-k). In contrast, *wt* RILP and ATG5 were coincident along the isolation membranes (Fig 14 b). RILP, thus, appears to be present at both nascent and more mature stages of isolation membrane formation in neurons. In contrast, GFP alone showed a diffuse cytoplasmic distribution (Fig 14 a, Fig 15 j), and had no detectable effect on DFCP1 or ATG5 distribution.

To address the molecular basis for RILP recruitment to isolation membranes, we examined the ability of our N- and C-terminal RILP fragments as well as the LIR3-mutant RILP to localize to these structures in cortical neurons expressing mCh-ATG5. A high colocalization was observed for both the N-terminal and LIR3-mutant RILP constructs, but almost none for the RILP C-terminal fragment alone (Fig 14 c-e, Fig 15 c-e). In the case of N-terminal RILP expression, a striking increase in crescent-shaped ATG5-positive isolation membrane profiles was observed (15.1% wt RILP *vs* 57.7% N-terminal RILP) (Fig 14 c, Fig 15 c). This suggests an arrest in maturation of these structures. Similar structures were also observed upon expression of the LIR3-mutant RILP (35.9% LIR3- mutant RILP) (Fig 14 e, Fig 15 e). In both cases, there was a corresponding decrease in the mature isolation membranes (Fig 14 i). Wild type full-length RILP, in contrast, exhibited a high degree of localization to more mature, vesicular isolation membrane profiles (59.1% full vesicles with *wt* RILP) (Fig 14 f-i). These data suggest that the RILP N-terminal domain is sufficient for its recruitment to isolation membranes, whereas the C-terminal domain is required for extension and successful closure of these structures.

The localization of the LIR3-mutant RILP to isolation membranes is presumably *via* the RILP N-terminal ATG5-binding site. These results also argue that LC3 is not the primary anchor for *wt* RILP on isolation membranes. This conclusion is supported by the failure of the RILP C-terminal fragment alone, which includes LC3- and Rab7-binding sites, to localize to isolation membranes (Fig 14 d). Together, these results suggest that RILP binds to a distinct anchor protein on isolation membranes through an N-terminal site, independent of LC3 and Rab7. We further found that RILP C-terminus showed distinct co-localization with both, mCh-Rab7 positive LEs and RFP-LC3 positive APs in glioma cells, which is consistent with an active role for the C-terminal Rab7- and LC3- binding sites in RILP (Fig 15 f, g). Conversely, the RILP N-terminal fragment, which contains dynein and ATG5-binding sites, showed a diffuse cytoplasmic distribution and no detectable recruitment to the LE or AP membranes (Fig 15 h, i).

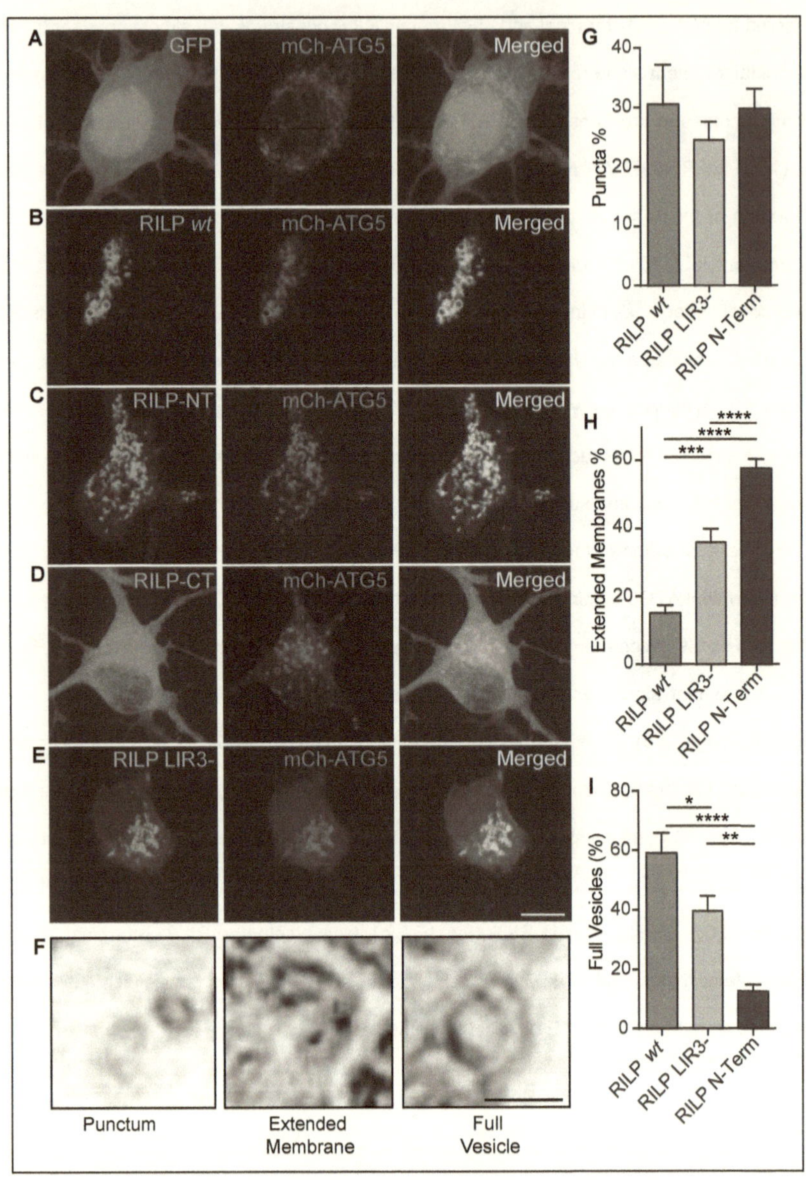

A
GFP
mCh-ATG5
Merged

B
RILP *wt*
mCh-ATG5
Merged

C
RILP-NT
mCh-ATG5
Merged

D
RILP-CT
mCh-ATG5
Merged

E
RILP LIR3-
mCh-ATG5
Merged

F
Punctum
Extended Membrane
Full Vesicle

G
Puncta %
RILP *wt*
RILP LIR3-
RILP N-Term

H
Extended Membranes %
***
****
****
RILP *wt*
RILP LIR3-
RILP N-Term

I
Full Vesicles (%)
*
****
**
RILP *wt*
RILP LIR3-
RILP N-Term

Figure 14. RILP is Recruited to Isolation Membranes in Neurons.
(a-e) GFP-tagged RILP, RILP mutants, and sub-fragments were co-expressed with ATG5, a marker for isolation membranes, in DIV7-8 cortical neurons. **(a)** GFP control vector shows diffuse cytoplasmic distribution unaffected by mCh-ATG5 expression. **(b)** GFP-RILP, **(c)** N-terminal RILP (1-185aa) and **(e)** LIR3- mutant RILP show specific localization to isolation membranes. **(d)** C-terminal RILP (210-401aa) shows no detectable localization to isolation membranes (n>20) (Scale bar: 5 µm) **(f)** Three distinct classes of ATG5 and RILP dual-positive structures were observed using live spinning disk confocal microscopy at 200X magnification: solid puncta, extended membranes and full vesicles (Scale bar: 1 µm). **(g)** Comparable number of ATG5 puncta observed in *wt*, LIR3- mutant and N-terminal RILP fragment expressing neurons. **(h)** Striking increase in extended ATG5 membranes in LIR3- mutant and N-terminal RILP expressing neurons. **(i)** Striking decrease in fully-formed ATG5 structures in LIR3- mutant and N-terminal RILP expressing neurons. P-values: *=0.05, **=0.01 ***=0.001, ****= 0.0001.

To investigate the association of RILP with ATG5-positive isolation membranes further, we tested for a potential RILP-ATG5 interaction biochemically. We found clear RILP co-immunoprecipitation with endogenous ATG5 in glioma cell extracts (Fig 17 a). Furthermore, recombinant bacterially-expressed full length or N-terminal RILP pulled down purified recombinant ATG5 supporting a direct RILP-ATG5 interaction through the RILP N-terminal domain (Fig 17 b). His-ATG5 also efficiently pulled down GST-tagged full length and N-terminal RILP (Fig 16 a, b), as well as the LIR3-mutant RILP (Fig 16 c, d). These results further indicated that the RILP LIR motifs are dispensable for the interaction with ATG5. Conversely, ATG5 failed to pull down the RILP C-terminal domain alone, which contains the Rab7- and LC3-binding sites (Fig 16 c, d). Together, our findings show that LC3 and Rab7 do not serve as anchors for RILP recruitment to isolation membranes. Instead, we find the well-established isolation membrane component, ATG5, which is essential for autophagosome biogenesis, to function as a critical RILP interactor, serving to recruit RILP to growing isolation membranes.

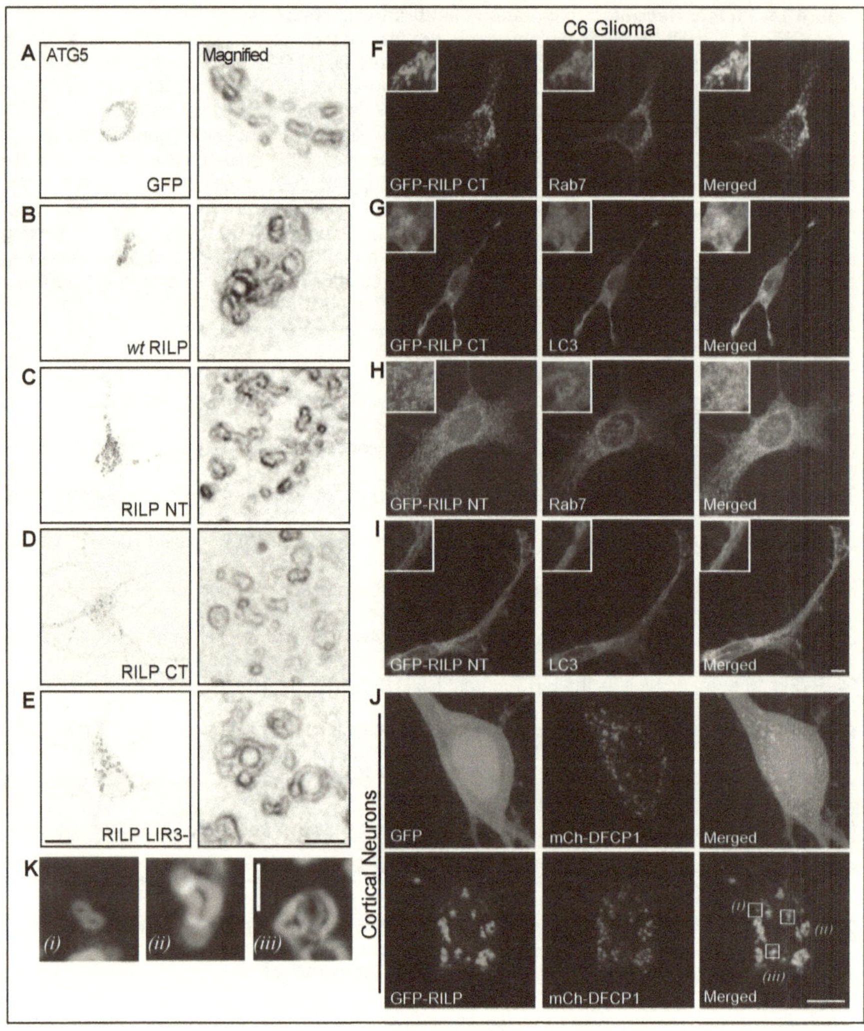

Figure 15. ATG5-positive Isolation Membrane Morphology in RILP Variants Expressing Neurons.
Magnified view of GFP-tagged RILP, RILP mutants, and sub-fragments co-expressed with ATG5, a marker for isolation membrane, in live DIV7-8 rat cortical neurons imaged with spinning disk confocal microscopy at 200X magnification, depicted in main figure 4. **(A-E)** Neuronal distribution and magnified view of the ATG5 structures are depicted for **(A)** GFP control vector **(B)** GFP-RILP **(C)** N-terminal RILP (1-185aa), **(D)** C-terminal RILP (210-401aa) and **(E)** LIR3-mutant RILP

expressing neurons. **(B)** *wt* RILP -positive isolation membranes appear as mature, vesicular structures. **(C** and **E)** RILP N-terminal domain or LIR3-mutant -positive isolation membranes appear as incomplete, extended structures, without apparent closure. **(D)** RILP C-terminus does not localize to isolation membranes. Scale bars: 5 µm for low and 2 µm for high magnification. GFP-tagged RILP C-terminus, containing Rab7- and LC3- binding sites, colocalizes with mCh-Rab7 **(F)** and RFP-LC3 **(G)** vesicles. GFP-tagged RILP N-terminus, containing dynein- and the newly identified ATG5- binding site shows diffuse cytoplasmic distribution and no recruitment to mch-Rab7 **(H)** or RFP-LC3 **(I)** positive structures (scale bar: 5µm). **(J)** RILP colocalization with mCh-DFCP1-labeled nucleation sites was monitored by dual-channel high-resolution live imaging of DIV7-8 rat cortical neurons. GFP control vector shows diffuse cytoplasmic distribution unaffected by mCh-DFCP1 expression. Full-length GFP-RILP staining of membrane structures juxtaposed with DFCP1-labeled nucleation sites often exhibited incomplete semi-circular pattern of growing isolation membranes (scale bar: 5 µm). **(K)** Classes of RILP structures observed juxtaposed with DFCP1: *(i)* puncta, *(ii)* crescent membranes and *(iii)* mature vesicles (scale bar: 1 µm).

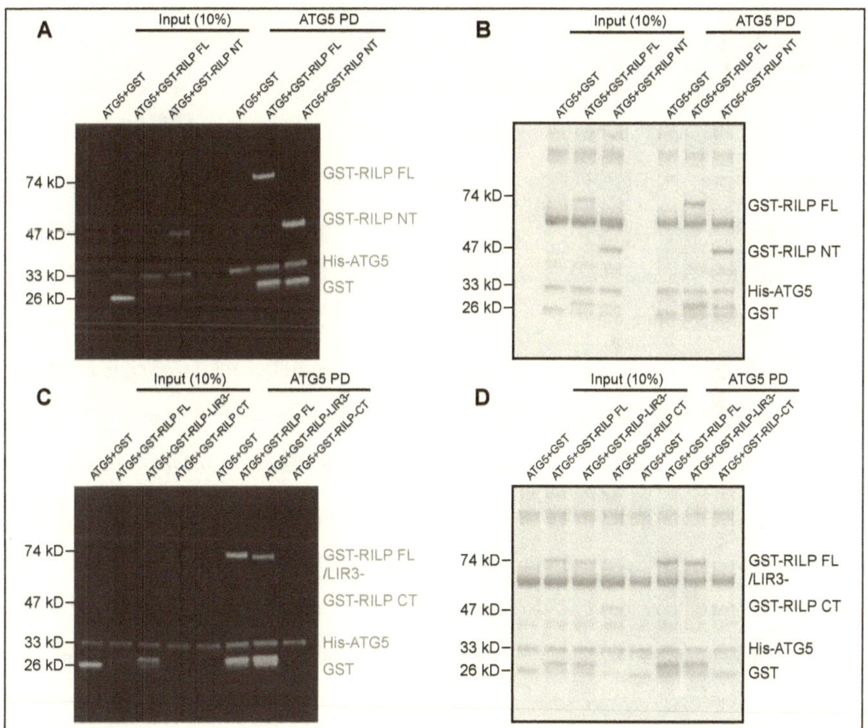

Figure 16. RILP Binds ATG5 in vitro.
To characterize the molecular interaction between RILP and ATG5, we performed *in vitro* direct
binding assays between GST-tagged versions of *wt*, LIR3-mutant RILP, RILP fragments and
His-ATG5. Immunoblotting **(A)** and coomassie staining **(B)** shows that His-ATG5 efficiently pulls
down GST-tagged full length *wt* RILP and RILP N-terminus, but not GST alone. Immunoblotting
(C) and coomassie staining **(D)** show that His-ATG5 also pulls down GST-tagged LIR3-mutant
RILP but not RILP C-terminus.

2.6. ATG5-mediated RILP Recruitment to Isolation Membranes is Independent of Dynein

In striking contrast to the highly motile RILP-positive autophagosomes, live imaging of

RILP-labeled isolation membranes revealed them to be mostly immotile (Fig 17 c). As noted,

ATG5 interacts with the N-terminal domain of RILP (Fig 17 b), which also has a well-established

role in dynein binding (Scherer et al., 2014). To compare the relative localization of dynein *vs*

ATG5 on isolation membranes, we directly tested for the presence of dynein on isolation

membranes using an antibody to the dynein LIC1 (Light Intermediate Chain 1) subunit. In DIV8

neurons co-expressing GFP-RILP and mCh-ATG5, we found that dynein was absent from RILP-

positive isolation membranes (87.2% APs). In contrast, dynein was enriched at ATG5-negative

fully-formed RILP-containing autophagosomes (69.4% APs) (Fig 17 d-e, quant. f-g). Thus, when

recruited to isolation membranes *via* ATG5, RILP fails to bind dynein. An *in vitro* competitive

binding assay between RILP and ATG5 *vs* dynein LIC1 showed that RILP binding to ATG5 was

strikingly decreased in the presence of LIC1 (Fig 18 a). In contrast, LC3 binding to RILP was

unaffected by ATG5. These data suggest that, although the dynein LIC1 and ATG5 may compete

for binding to RILP, LC3 and ATG5 interact with RILP independently (Fig 18 b). We did note a

small population of RILP–positive structures in cortical neurons that were enriched for both, ATG5

and dynein (4.4% triple positive vesicles), which might represent a transition from ATG5- to

dynein-bound structures. In view of the transition of APs from immotile to motile structures, we

speculate that this is likely controlled by dynein displacement of ATG5 on RILP.

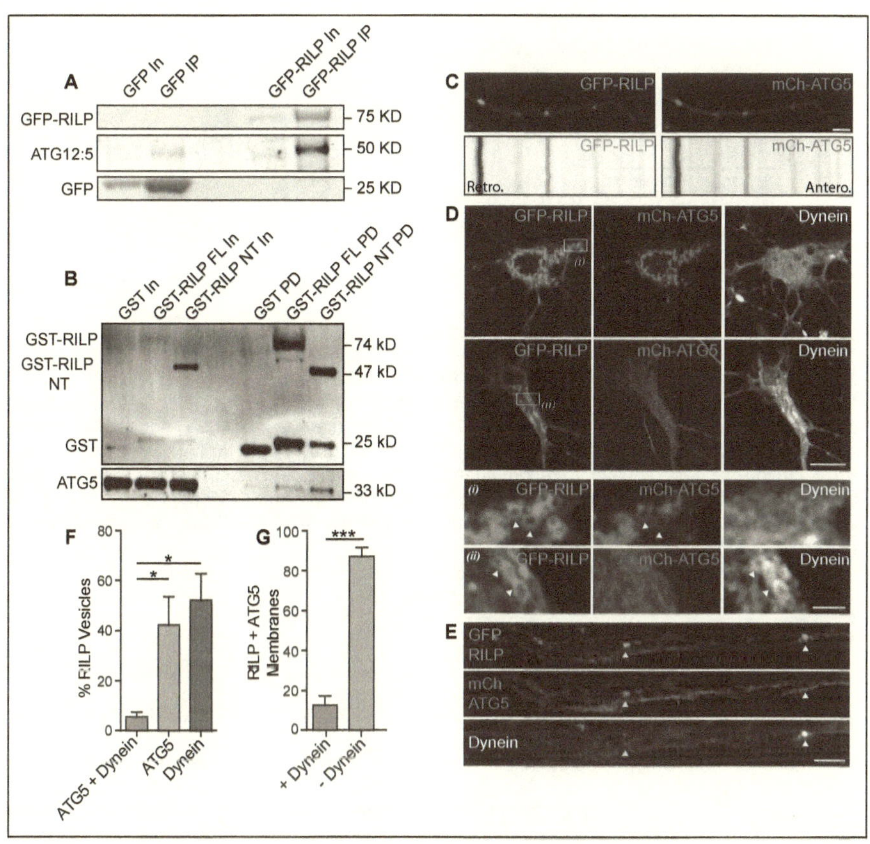

Figure 17. ATG5 Directly Binds RILP and Prevents Dynein Recruitment to Isolation Membranes.

(A) Full-length GFP-RILP immunoprecipitated endogenous ATG5:12 from C6 glioma cell lysates, visualized by an anti-ATG5 antibody. **(B)** Full- length and N-terminal GST-RILP pulled down recombinant His- ATG5 *in vitro*. Live cell time-lapse imaging was used to monitor GFP-RILP behavior at mCh-ATG5-labeled isolation membranes in DIV7-8 cortical neurons. **(C)** Single frames for individual axons are shown at top for *wt* GFP-RILP and mCh-ATG5. Shown below are kymographs which indicate the positions of individual isolation membranes over time. Scale bars: x = 5µm; y= 30sec. RILP-positive isolation membranes were mostly immotile. **(D)** Immunostaining for dynein LIC1 subunit showed that ATG5-positive RILP structures were devoid of dynein (Top, (*i*)), while ATG5-negative RILP structures were enriched for dynein (bottom, (*ii*)) in soma as well as **(E)** neurites. **(F)** Percent relative localization of RILP with ATG5 and/ or dynein. **(G)** Majority of RILP-positive isolation membranes are devoid of dynein. Scale bars: 5 µm. p-values: *=0.05, ***= 0.001.

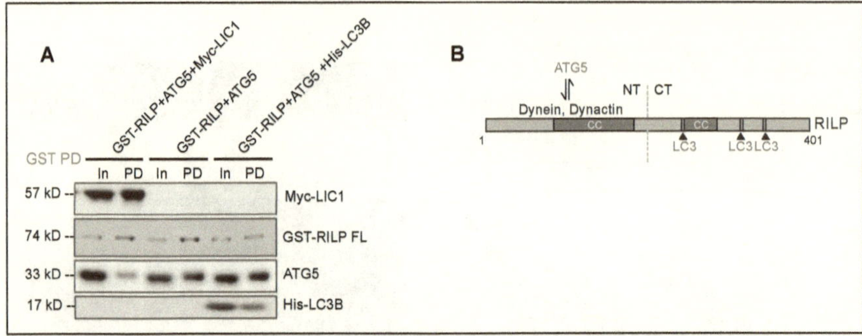

Figure 18. ATG5 and Dynein Compete for RILP Binding.
(A) To determine whether ATG5 and dynein compete for a common binding site in RILP, we performed an *in vitro* competition assay. Addition of Myc-tagged dynein LIC1, reduced the amount of ATG5 pulled down by GST-tagged *wt* RILP, suggesting that LIC1 may displace ATG5 from RILP. On the contrary, addition of recombinant LC3 did not alter the amount of ATG5 pulled down by RILP, suggesting a simultaneous binding to RILP of ATG5 and LC3. **(B)** As depicted in the diagram, these biochemical data suggest that ATG5 binds within the N-terminal dynein-binding domain of RILP.

2.7. Discussion

Autophagy plays fundamental roles in yeast and mammalian homeostasis. Neuronal autophagy is of particular interest for its involvement in nervous system development and neurodegeneration among other roles (Cecconi & Levine, 2008; Sarkar, Ravikumar, & Rubinsztein, 2009; G. Tang et al., 2014) . However, the molecular mechanisms mediating and regulating neuronal autophagy have remained poorly understood. Although the core yeast autophagy machinery is conserved in mammals, additional mechanisms might have evolved to regulate autophagosome behavior in the morphologically more complex, highly compartmentalized and long-lived mammalian cells, especially neurons. Our study now identifies a neuronal autophagy mechanism involving the dynein adaptor protein RILP. Surprisingly, we find that RILP performs multiple distinct functions during autophagosome biogenesis and subsequent retrograde transport, and is essential for autophagic clearance. Interestingly, we also find these RILP functions to be responsive to induction of autophagy by mTOR inhibition.

RILP Interaction with Autophagosomes through LC3

We have found that the dynein adaptor protein RILP associates with two core autophagosomal membrane proteins, ATG5 and LC3, through distinct N- and C-terminal binding sites, respectively. RILP showed extensive colocalization and co-transport with LC3-positive autophagosomes in cortical neurons. This behavior was severely inhibited by RILP depletion, supporting the physiological significance of the RILP-LC3 interaction. We identified three LC3-binding (LIR) motifs within the RILP C-terminal domain. Mutational analysis of these motifs showed that all three are functional and necessary for LC3- binding, RILP recruitment to neuronal APs, and their retrograde transport. In contrast, the previously identified RILP Rab7-binding site was dispensable for RILP recruitment to APs, though essential for its recruitment to LEs. RILP's dual interaction with LC3 and Rab7 suggests that this protein may associate with and transport both early (LC3-positive) and late (LC3-Rab7 dual-positive) autophagosomal structures in neurons (Figure 19 *c-e*).

RILP Control of Autophagosome Biogenesis

Surprisingly, the RILP LIR mutations resulted in a striking decrease not only in AP transport, but also in autophagosome number, an effect not observed for the Rab7-binding mutation. RILP knockdown also showed a similar phenotype, supporting an unexpected and independent RILP role in AP formation. Our data suggest that RILP may affect AP biogenesis at the stage of isolation membrane maturation and/or closure. Our colocalization analysis revealed that RILP associates with early, pre-closure autophagosomal intermediates in neurons, including DFCP1-positive nucleation sites as well as ATG5-positive isolation membranes (Fig 19 *a, b*). RILP was observed to be slightly off-set from DFCP1 at the nucleation sites, but more clearly co-localizing with the nascent isolation membranes extending from these sites. RILP and ATG5, on the contrary, exhibited clear coincidence in distribution along the isolation membrane structure.

We found further, as a basis for this behavior, physical interaction between RILP and ATG5 *in vitro* as well as in cell lysates. Unlike LC3 and Rab7, however, ATG5 bound to the RILP N-terminal domain consistent with distinct functional roles for the ATG5 *vs* LC3 and Rab7 interactions. Although the RILP N-terminal domain was sufficient for recruitment to isolation membranes, it was not sufficient for their maturation into fully-formed autophagosomes. Expression of the RILP N-terminal domain alone, which lacked the three LIR motifs, caused a notable change in isolation membrane morphology, resulting in an apparent extension of the isolation membranes without closure. A similar defect was observed when the full length LIR3-mutant RILP was expressed in neurons. Thus, we speculate that the RILP-LC3 interaction may play an additional critical role in the proper maturation of isolation membranes into autophagosomes. Our data are consistent with a model according to which RILP would first become anchored to the isolation membrane through ATG5. The RILP C-terminal LIR motifs might then serve to accumulate LC3 molecules for their lipidation and incorporation into the growing isolation membranes by components of the ATG5-12:16 complex.

RILP Coordination of Autophagosome Formation with Motility

Using high temporal resolution live imaging, we found that RILP contributes to axonal transport of LC3- positive autophagosomes in the retrograde direction. Although RILP can be detected on isolation membranes, these structures are immotile, as demonstrated here by live imaging analysis. Direct *in situ* analysis of dynein distribution showed an absence of the motor protein from ATG5-positive isolation membranes, but a clear association with ATG5-negative RILP-containing structures. These observations suggest an unusual mechanism for RILP activation of dynein-mediated transport in concert with completion of isolation membrane closure and departure of the ATG12:5-16 complex.

As RILP binds to both ATG5 and dynein through the N-terminal domain, but rarely colocalizes with both markers, we propose that RILP function at the isolation membrane is dynein-independent. Our data suggest that RILP is initially recruited to the isolation membranes through its N-terminal interaction with ATG5. Dynein is detected predominantly on ATG5 negative membranes, suggesting that dynein may displace ATG5 from a common RILP N- terminal binding site. At this stage, RILP C-terminal LIR motifs may facilitate LC3 sequestration close to growing isolation membrane. Indeed, *in vitro* binding experiments show that RILP is able to simultaneously bind both ATG5 and LC3 and, unlike dynein, the presence of LC3 does not affect the efficiency of RILP binding to ATG5. Upon isolation membrane closure and departure of the ATG5:12-16 complex (Carlsson and Simonsen, 2015), RILP, through its C-terminal LIR motifs, may remain associated with the newly recruited LC3 on the autophagosomes. At this stage, we observe that RILP recruits dynein and confers motility to the fully-formed APs. Thus, the sequential RILP interactions with ATG5, LC3 and dynein that we identify here would serve to coordinate isolation membrane extension with the start of retrograde autophagosome motility. This would ensure that, wherever within the cell autophagic events occur, the completed AP should immediately begin its journey to the cell body for degradation.

mTOR Regulation of RILP Function and Autophagic Cargo Clearance

Autophagy is negatively regulated by mTOR kinase in yeast and non-neuronal mammalian cells. However, a functional link between mTOR activity and neuronal autophagosome behavior has remained incompletely explored. Perturbations of mTOR function have been implicated in several neurodevelopmental processes as well as neurodegenerative disorders (Abe, Borson, Gambello, Wang, & Cavalli, 2010; Crino, 2013; Kwon et al., 2006; Nie et al., 2010; G. Tang et al., 2014). Despite this, the underlying cellular and molecular mechanisms, including the contributions of autophagy, remain to be fully understood.

We find that induction of autophagy by mTOR inhibition results in transcriptional and translational upregulation of RILP, but not of several other known dynein or kinesin adaptors. Under these conditions, we observe an increase in RILP mRNA as well as protein levels and a clear increase in RILP-positive vesicles in neuronal and non-neuronal cells. Our data suggest, therefore, that upon mTOR inhibition RILP may be induced in order to participate in autophagosome formation and transport. Previous work has shown that upon mTOR inhibition, LAMP1-positive vesicles in non-neuronal cells show a redistribution from cell periphery towards the nucleus. Our data now suggest that this redistribution might also be under RILP control, though this possibility remains to be explored. Conceivably, RILP expression or function might also be dysregulated in neuropathological disorders with known mTOR dysfunction, which in turn could affect the RILP-mediated autophagy pathway. Evidence for such a role is so far lacking, and further work is likely needed to test this possibility more directly.

Subcellular aggregation of p62 has been reported in several neurodegenerative disorders, but the molecular mechanisms leading to this pathology have remained incompletely understood. We find that RILP depletion or loss of its LC3-binding capacity potently interfere with autophagic clearance of Sequestosome1/ p62, leading to its cytoplasmic aggregation. The p62 aggregates are enriched for ubiquitin as well as LC3, but are mostly negative for Rab7. This suggests that RILP depletion arrests p62 turnover during the early stages of the autophagy pathway, prior to its delivery to the Rab7-positive late endosomal compartment. Mutations of the three RILP LIR motifs also inhibit p62 turnover, which suggests that RILP's association with autophagosomes, in particular, is required in this function. Furthermore, it has been reported that apart from binding to LC3, the LIR motif in p62 may play a role in establishing a crosstalk between the ubiquitinated substrates and the autophagy machinery to assist AP formation (Danieli & Martens, 2018; Zaffagnini et al., 2018a). Whether RILP can potentially play a role in stabilizing this crosstalk and in assisting cargo capture by the autophagy machinery acting in concert or independently of p62, remains to be tested. Overall, our data lead to the conclusion that RILP is required for autophagic clearance, presenting the interesting possibility that enhancement of RILP function, for example through mTOR inhibition, might help alleviate neurodegenerative pathology.

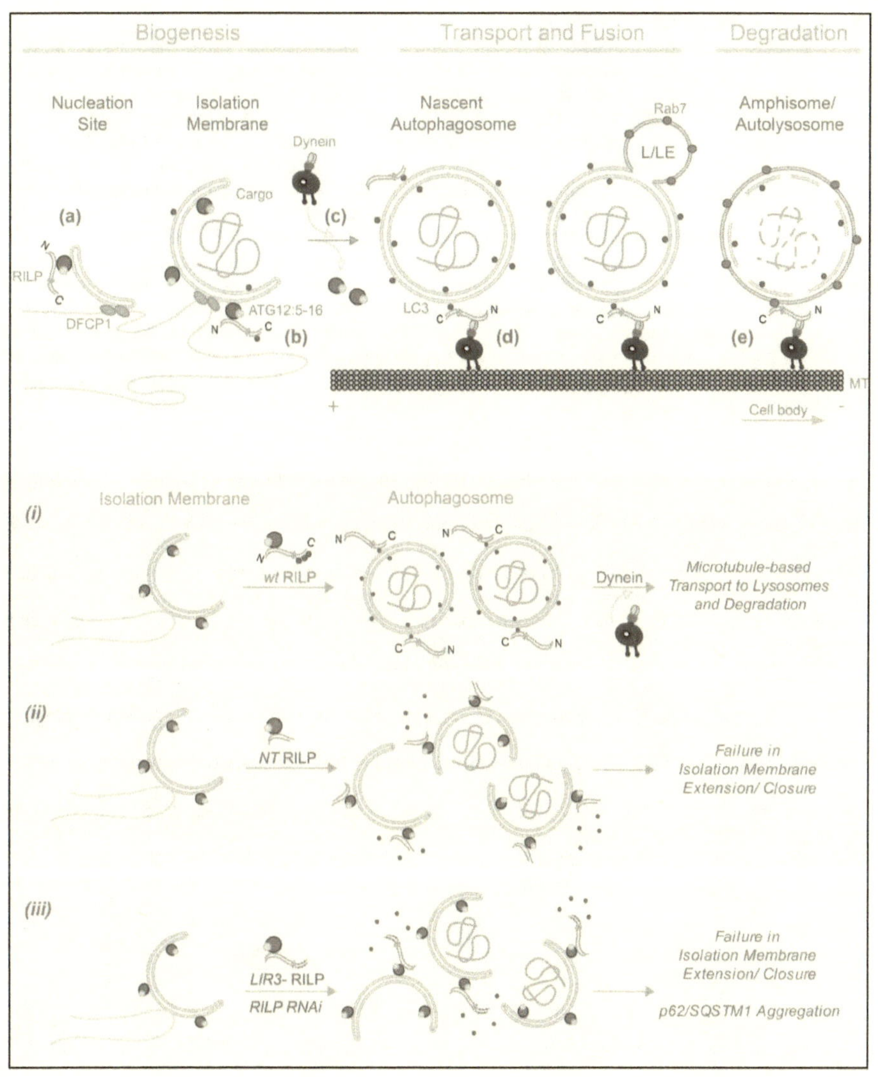

Biogenesis
Transport and Fusion
Degradation
Nucleation Site
Isolation Membrane
Nascent Autophagosome
Amphisome/ Autolysosome
Dynein
Rab7
L/LE
(a)
Cargo
RILP
N
C
DFCP1
ATG12:5-16
N
C
(b)
(c)
LC3
C
N
(d)
C
N
C
N
(e)
MT
+
Cell body
-
(i)
Isolation Membrane
Autophagosome
N
C
N
C
wt RILP
Dynein
Microtubule-based Transport to Lysosomes and Degradation
C
N
C
N
(ii)
NT RILP
Failure in Isolation Membrane Extension/ Closure
(iii)
LIR3- RILP
RILP RNAi
Failure in Isolation Membrane Extension/ Closure
p62/SQSTM1 Aggregation

Figure 19. A Model for RILP Mediated Autophagy in Neurons.
Stages in autophagy progression are shown, along with stage-specific protein markers, RILP, and dynein associated with the microtubule at the bottom. Nucleation, expansion of isolation membrane and maturation of the autophagosome are depicted. RILP shown as an elongated coiled coil-containing dimer, is first detected in juxtaposition with nucleation sites *(a)*, but concentrated at nascent and more fully formed isolation membranes, associated with ATG5 *(b)*. RILP then appears at mature autophagosomes, linked through LC3 *(c, d)*. RILP remains associated with AP-LE membranes through LC3 and Rab7 *(e)*, potentially until lysosomal fusion. This pathway is activated by mTOR regulation of RILP expression, and functions in autophagic clearance. *(i-iii)* Effects of RILP perturbations on autophagy pathway: *(i)* Full-length *wt* RILP binds ATG5 on isolation membranes through N-terminus and may recruit or stabilize LC3 at these structures through the C-terminal LIR motifs. This allows for normal autophagosome formation and eventual retrograde transport to lysosomes. *(ii)* RILP N-terminal domain is recruited to ATG5 on isolation membranes, but fails to bind LC3. This results in an arrest in the maturation of isolation membranes into fully-formed autophagosomes. *(iii)* RILP LIR-mutant expression or RNAi also result in an arrest in the maturation of isolation membranes and an aggregation of p62/ Sequestosome-1 in cells.

Together our results strongly support a new model for neuronal autophagy, according to which dynein adaptor RILP coordinates neuronal autophagosome biogenesis and transport to maintain neuronal homeostasis. The ability of RILP to associate with both growing isolation membranes and fully-formed autophagosomes, as well as Rab7-positive post-fusion autophagosomal membranes identifies a highly unusual range of interactions within a common pathway. Conceivably, RILP expression or function might be deregulated in neuropathological disorders, especially those known to have an aberrant mTOR activity. Further insights into RILP behavior in these conditions would help identify roles for RILP-mediated autophagy in neurodevelopmental and degenerative pathways, with potential therapeutic implications.

CHAPTER 3. PHYSICAL AND PHYSIOLOGICAL FUNCTIONS OF RILP

RILP associates with dynein and its regulator dynactin to form a multi-subunit complex that transports cellular cargoes toward the minus ends of microtubules. My biochemical analysis of the composition of the RILP supercomplex reveals that it contains LIS1, a dynein regulator implicated in *lissencephaly* but not NudE, another regulator implicated in *microcephaly*. Unlike other well-studied dynein adaptors such as BicD2 and its homologues, RILP is able to precipitate high levels of endogenous LIS1 along with dynein, suggesting a strong LIS1 affinity for RILP supercomplex and a potential obligate role for LIS1 in RILP behavior *in vivo*. In this chapter, I discuss the composition of the RILP-dynein supercomplex, the biophysical properties of this complex, and their implications for cellular functions of RILP in autophagosome and endosome transport.

Secondly, the novel roles for RILP that we identified in *Chapter 2* raise questions regarding the expression and behavior of RILP in pathophysiological conditions, particularly those concerning neurodevelopment and degeneration. My preliminary studies suggest that indeed, the expression of RILP is altered during neurodevelopment and might be affected during neurodegeneration, suggesting that the pathway we identified might have broad implications *in vivo*.

3.1. Composition of the RILP-dynein supercomplex

Previous studies on adaptor proteins including BicD2, BicDR1 and HOOK3 have shown, that these adaptor proteins can regulate dynein force production, run lengths and velocities, as well as the number of motors recruited to cargoes. Thus, we sought to define the molecular composition of the RILP-dynein complex, in order to characterize its biophysical properties.

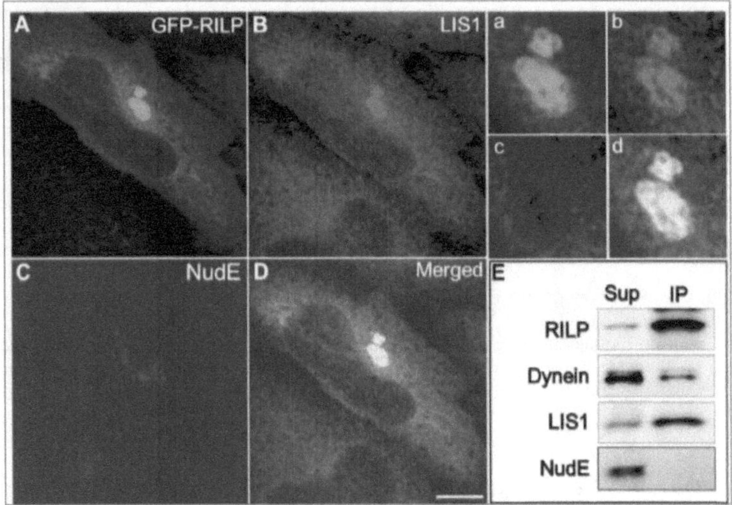

Figure 20. RILP recruits LIS1 but not NudE to Dynein Supercomplex.
(A) GFP-tagged RILP, when expressed in HeLaM cells, shows a pericentrosomal (microtubule minus-end) clustering of vesicles. Immunofluorescence analysis shows these vesicles to be enriched for (B) Lis1 but not for (C) NudE. (D) Merged image from A-C. (a-d) A higher magnification view shows that RILP and LIS1 are present on vesicular membranes but not the lumen. (E) Endogenous RILP immunoprecipitates dynein (intermediate chain 74.1 is blotted for in this figure) and LIS1, but not NudE, consistent with our immunofluorescence analysis.

For this purpose, we first expressed GFP-tagged full length RILP in HeLaM cells and performed an anti-GFP immunoprecipitation to probe for subunits of dynein, dynactin p150[glued] and any other dynein regulators, such as LIS1 and NudE/EL. We found that GFP-tagged full length RILP efficiently co-immunoprecipitated dynein IC, HC as well as dynactin p150^{glued}.

Interestingly, RILP also pulled down striking levels of LIS1, but no NudE/EL (Figure 20 E).

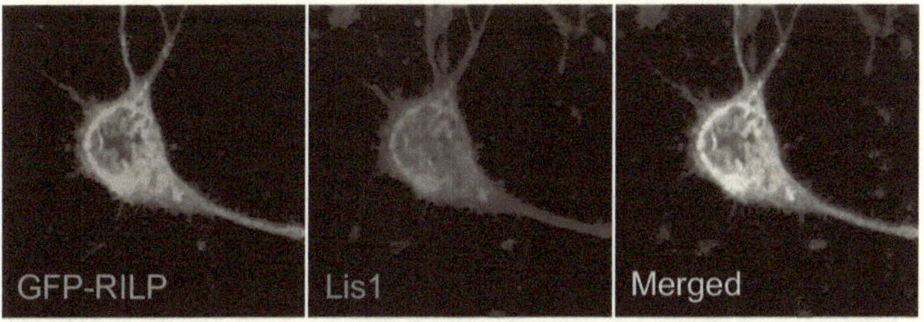

Figure 21. RILP Recruits Endogenous LIS1 to Vesicular Membranes in Hippocampal Neurons.
GFP-RILP positive vesicles in Hippocampal DIV3 neurons are enriched for endogenous LIS1, consistent with our non-neuronal observations.

We further tested for the presence of LIS1 and NudE/EL on RILP-positive vesicles in cells. We found that indeed, RILP vesicles are enriched for endogenous LIS1 in both, neuronal and non-neuronal cells (Figures 20 and 21). However, NudE/EL was strikingly absent from these membranes. Altogether, these results suggest that RILP forms a tetra-partite complex with LIS1, dynein and dynactin, but excludes NudE/EL.

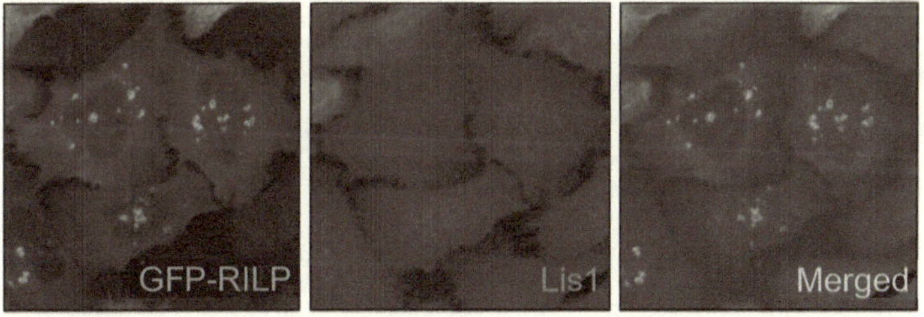

Figure 22. Lis1 is Required for RILP Mediated Retrograde Transport.
Dynein regulator Lis1 was depleted using siRNA pool in GFP-RILP expressing HeLaM cells. GFP-RILP positive vesicles showed cytoplasmic dispersal in the absence of Lis1, suggesting that Lis1 is essential for RILP-mediated retrograde motility. Depletion of Lis1 in RILP expressing cells was confirmed using anti- Lis1 antibody.

Next, we wanted to test whether consistent with our immunofluorescence and biochemical data, LIS1 is indeed required for RILP mediated motility of cargoes, and whether NudE/EL were dispensable for the same. For this purpose, we first transfected HeLaM cells with either NudE or LIS1 siRNAs on Day 1. After 24 hrs of siRNA treatment, we transfected full length GFP-RILP in these cells, which were fixed after another 24 Hrs of incubation on Day 3 and

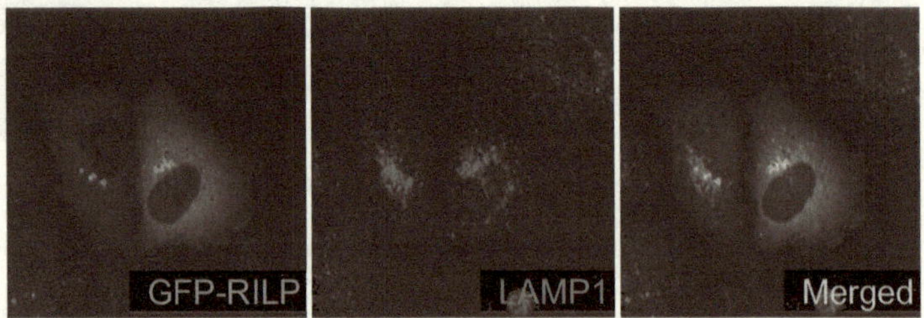

Figure 23. NudE is not Required for RILP Mediated Retrograde Transport.
Dynein regulator NudE was depleted using siRNA pool in GFP-RILP expressing HeLaM cells. GFP-RILP positive vesicles show perinuclear accumulation even in the absence of NudE, suggesting that NudE is not required for RILP-mediated retrograde motility. LAMP1 positive lysosomes in these cells also show a slight accumulation at cell center, suggesting that they might also be under control of RILP.

stained for endogenous LIS1 and NudE/EL as well as GFP-RILP. We found that in NudE KD cells, RILP-positive vesicles were clustered closer to the centrosome, very similar to that in control cells (Figure 23). However, in LIS1 KD cells, RILP vesicles were mostly dispersed in the cytoplasm, suggesting that depletion of LIS1 interfered with RILP's ability to drive the retrograde vesicular transport. This indicates that LIS1 may be critical for RILP-mediated retrograde transport in cells, while NudE/EL is dispensable. The implications of these observations for RILP regulation of dynein motor behavior and for RILP control of autophagy are discussed in *Chapter 4*.

3.2. mTOR Regulation of RILP *in vivo*

To test whether mTOR-responsive upregulation of RILP that we saw in cultured neuronal and non-neuronal cells is physiologically relevant and occurs *in vivo*, we probed for RILP protein levels in postnatal brain striatal tissue of systemically rapamycin treated mice. We found that

rapamycin-treated striatal tissue showed a significant (~0.6-fold) increase in RILP protein levels,

compared to mock treated tissue (Figure 24, n=3). This observation is consistent with our cellular

studies. Thus, mTOR inhibition *in vivo* also results in an increased RILP expression in the

striatum, the region of the brain affected in ASD patients. Implications of this observation for RILP

roles in neurodevelopment are discussed in *Chapter 4*.

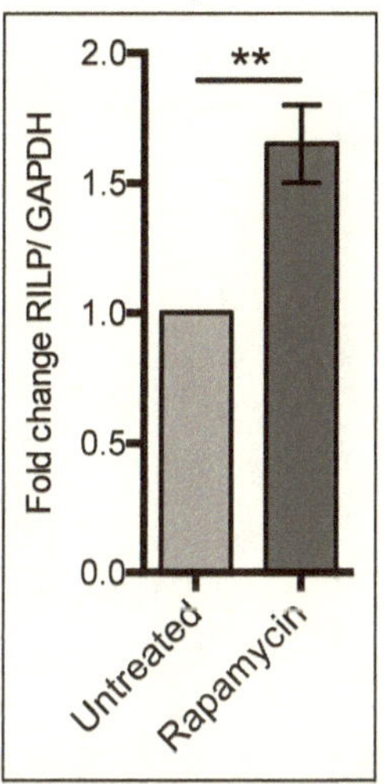

Figure 24. RILP is Upregulated Upon Rapamycin Treatment.
Mice were systemically treated with rapamycin from postnatal day 21 to postnatal day 28, sacrificed, and striatal tissue harvested for western blotting. Compared to control mock treated mice, Rapamycin treated cohort showed a significant increase in the levels of RILP protein, suggesting that consistent with our *in vitro* cellular data, RILP is upregulated upon mTOR kinase inhibition *in vivo*. *(Tissue samples kindly provided by Ori J. Leiberman, Sulzer lab, CUIMC)*

3.3. RILP Roles in Amyloid Beta Aggregate Clearance in Alzheimer's Disease

My recent work shows that RILP-mediated autophagy is required for p62/Sequestosome-

1 turnover in cells. Cytoplasmic p62 aggregates are a key feature of neurodegenerative pathology

in Alzheimer's, Huntington's and Amyotrophic Lateral Sclerosis. Molecular basis for p62

aggregation in these diseases is not well understood and has been attributed to defects in AP/LE function (Liu & Li, 2019), or to changes in p62 expression (Caccamo, Ferreira, Branca, & Oddo, 2017a), although the exact molecular mechanisms that go awry leading to p62 aggregation remain to be identified. It is possible that RILP upregulation in neurodegeneration may enhance aggregate clearance and restore AP-LE function.

In Alzheimer's Disease brain tissue, an accumulation of immature autophagic vesicles has been reported in dystrophic neurites in the brain before the appearance of NFTs and amyloid beta plaques (R. A. Nixon et al., 2005; W. H. Yu et al., 2005).

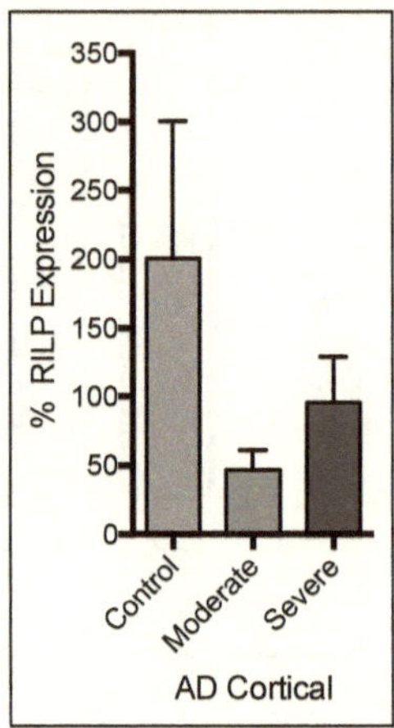

Figure 25. RILP Expression in Alzheimer's Disease Brain Tissue.
RILP protein levels were observed using anti-RILP antibody in human post-mortem Alzheimer's Disease brain hippocampal tissue (normalized to GAPDH). Unlike control tissue, tissue from moderate AD patient brains showed a striking decrease in RILP protein levels (n=3 individuals/ stage of the disease). Moderate and severe stage AD tissue showed decreased RILP protein levels compared to age-matched control brains. Most striking decrease was observed in the moderately affected tissue, with a slight relative increase in severe stage *(Samples kindly provided by Lily Qu, Bartolini Lab)*.

The widely-studied AD mouse models (APP/PS1 and J20) each show p62 aggregation (Kuusisto, Salminen, & Alafuzoff, 2002). How autophagic dysfunction results in onset of AD pathology, however, remains poorly understood. RILP contributes to protein homeostasis and might well play a role in abnormal protein clearance or serve as a novel molecular tool to modulate maturation and turnover of AP/LEs in AD brains. In preliminary experiments, we find that RILP expression is high during early mouse brain development, but decreases ~4-fold in adulthood (Figure 26).

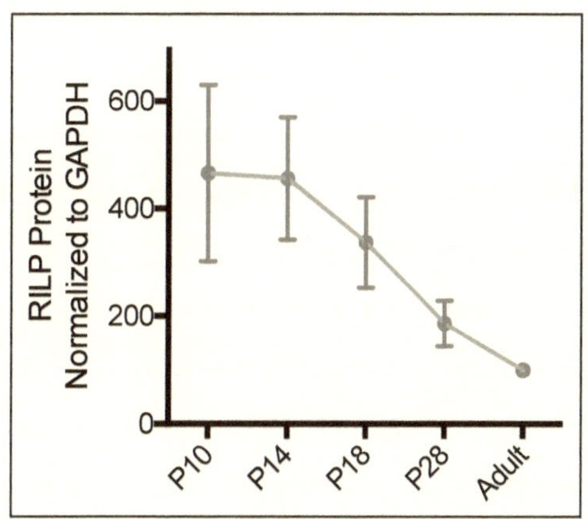

Figure 26. RILP Protein Expression During Striatal Development in Mice.
RILP protein expression drops almost 3.5- fold from early postnatal development (P10) to adulthood (3- month old mice).

In collaboration with Dr. F. Bartolini at Columbia University, we find RILP levels to be strikingly reduced in moderate and severe human AD brain tissue compared to age-matched controls (Figure 25). This reduction in RILP might well suppress AP/LE function and turnover, in turn resulting in AD pathology. Interestingly, RILP levels are upregulated in severe AD pathology suggesting that protein aggregation might upregulate RILP expression however, at later stages of this disease elevated RILP levels may not be effective for aggregate clearance. These observations suggest that early autophagic upregulation might aid in aggregate clearance and help prevent severe pathology. Further implications of this observation will be discussed in *Chapter 4.*

CHAPTER 4. FUTURE DIRECTIONS

The observations stated in this study provide several novel insights that re-define conventional roles for a 'dynein adaptor protein' RILP from a physical scaffold between dynein and cellular cargoes to a sophisticated multivalent interactor, *capable of organizing and regulating complex biological pathways*. The multiplicity of interactions invokes potential for a tight regulation of these interactions based on cell state, microenvironment and metabolic signaling, adding impressive layers of intra- and inter-molecular regulation.

Furthermore, the identification of a direct physical competition between dynein and ATG5 - the isolation membrane/ phagophore protein- to bind to a common RILP N-terminal domain shows RILP to function at *the interface of two complex cellular machineries*, namely, the autophagy machinery and the multi-subunit dynein motor complex. As such, RILP would have the ability to control biogenesis as well as retrograde transport of an autophagosome during its lifetime, thus exhibiting temporal control of the autophagy pathway.

The novel roles for RILP in cellular autophagy that we identified here raise a number of questions regarding the physiological implications of RILP malfunction or deregulation *in vivo*. There are, to our knowledge, no disease-associated human mutations so far found in RILP. This could be because of potential lethality of pathogenic mutations in this protein. Alternatively, it is possible that RILP might be deregulated through alterations in gene copy number and mRNA or protein expression, instead of sustaining pathogenic point mutations. Given mTOR regulation of RILP expression and behavior that we have observed, it is conceivable that diseases that exhibit abnormal mTOR behavior may show alterations in RILP function, in turn deregulating the autophagy-lysosomal pathway. Further investigations in this direction would provide valuable insights into disease pathology and may help test the therapeutic potential of RILP as a controller of the autophagy-lysosomal pathway *in vivo*.

Given the layers of complexity involved in RILP interactions with diverse cellular proteins and RILP regulation, five major directions of future work can be envisioned with the objective of understanding further RILP function at molecular as well as systemic levels.

4.1. Defining the ATG5-binding Site in RILP and Dissecting the Molecular Mechanism of Dynein-ATG5 competition.

We find that in DIV8 cortical neurons, dynein and ATG5 do not simultaneously co-localize with RILP, rather they tend to exhibit a mutually exclusive enrichment on RILP-positive structures.

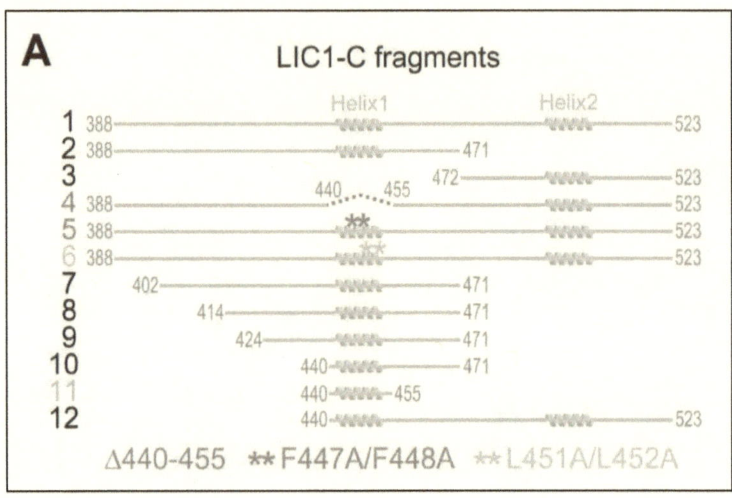

Figure 27. Mapping the minimal adaptor- binding LIC1 C-terminal fragment
A structure-function analysis was performed by Celestino et. al. with the above depicted fragments, including two mutated versions of LIC1 Helix1 sequence where consecutive phenylalanine and leucine residues were mutated to alanine residues. The authors identified differential binding sites within LIC1 C-terminus for different dynein adaptors, with RILP requiring minimal binding site within LIC1 (Adapted from Celestino et. al., 2019).

A detailed analysis of the novel RILP interaction with ATG5 shows that both full length and N-terminal dynein binding domains of RILP are efficiently recruited to ATG5-positive isolation membranes (i.e. phagophores). The C-terminal Rab7- and LC3-binding domain of RILP fails to localize to phagophores, and shows mostly diffuse cytoplasmic localization. In keeping with this observation, the LC3-binding deficient LIR3- mutant RILP efficiently co-localizes with ATG5 positive phagophores in neurons. These data indicate that RILP binding to ATG5 on isolation membranes requires the N-terminal dynein-binding domain in RILP, but not the LC3-binding sites.

A recent study published by Reto Gassmann's group (R. Celestino et al., 2019) has elegantly identified and compared the minimal binding site in dynein light intermediate chain 1 (LIC1) for several well-known dynein adaptor proteins. This work shows that the N-terminal region of dynein LIC1 and LIC2 contains a GTPase-like domain, which directly interacts with the N-terminal tail of the dynein heavy chain (the C-terminus of dynein heavy chain forms the motor domain). However, the C-terminal region (*aa 388-523*) of LIC1 and LIC2 is inherently disordered with a propensity for forming two very dynamic alpha-helical structures located at *aa 440-455* and *aa 497- 504*. *In vitro* binding analysis further shows that Helix 1 (*aa 440-455*) within LIC1 is sufficient for robust RILP binding. Unlike RILP, other dynein adaptor proteins including FIP1, BicD2, HOOK3, and Ninein require additional contact sites within LIC1 Helix 2 and a few conserved residues upstream of Helix 1 for LIC1 binding. Thus, RILP requires minimal contact within LIC1 C-terminal domain for a robust interaction, suggesting a high RILP binding affinity to dynein motor complex.

Our own *in vitro* biochemical analysis shows that full length as well as N-terminal RILP binds directly to ATG5. Addition of recombinant full length LIC1 to this reaction reduces the amount of ATG5 pulled down by RILP. This *in vitro* competition experiment shows that dynein LIC1 and ATG5 might compete for a common binding site in RILP.

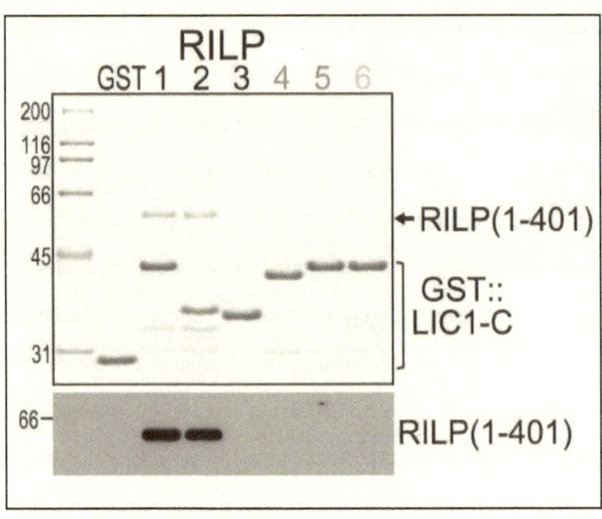

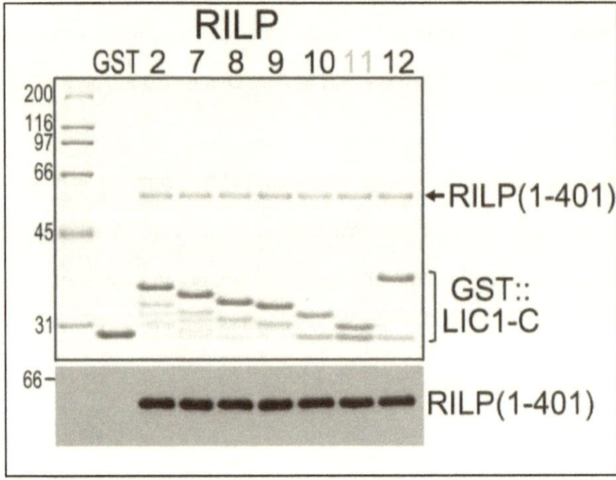

Figure 28. LIC1 C-terminal Helix 1 Region is Critical for RILP Binding.
Full length recombinant RILP protein is able to bind to LIC1 fragments that contain Helix 1 region, but fails to bind LIC1 fragments deleted for Helix1, and mutated to convert FF and LL motifs within Helix1 to AA. Altogether, these experiments show that RILP binding to LIC1 is dependent on LIC1 C-terminal Helix 1 domain (Adapted from Celestino et. al., 2019).

Now that the minimal RILP-binding region in LIC1 has been identified and experimentally validated, it will be interesting to further dissect the exact ATG5 binding site in RILP. This will in turn provide mechanistic insights into the regulation of RILP binding to ATG5 *vs* LIC1 during AP biogenesis and subsequent transport, respectively.

First, it will be important to test whether the C-terminal domain of LIC1 (*aa 338-553*) and specifically, Helix1 (*aa 440-455*) of LIC1 can robustly bind full length recombinant RILP *in vitro*. Further analysis can then be restricted to N-terminal RILP, which in our hands is sufficient in co-localizing with ATG5 *in vivo* and in pulling down ATG5 *in vitro*. The recombinant N-terminal RILP should be able to efficiently pull down LIC1 helix 1 fragment. A competition experiment to compare the relative quantities of ATG5 pulled down by N-terminal RILP in the presence/ absence of LIC1 Helix 1 may reveal that the presence of LIC1 Helix 1 decreases the amount of ATG5 pulled down by RILP N-terminus. This observation would suggest a direct competition between LIC1 Helix1 and ATG5 for binding RILP N-terminus.

It will be important to test whether the two mutated versions of LIC1 Helix 1 construct where two Phenylalanine residues (FF 447-448) or two Leucine residues (LL 451-452) are changed to Alanine (AA) can compete with ATG5 for RILP binding. Based on the Gassmann study, I predict that the FF or LL mutant LIC1 Helix 1 versions should fail to bind RILP N-terminus in a direct *in vitro* binding experiment, and thus fail to compete with ATG5 for RILP binding. Thus, the FF or LL mutant LIC1 Helix 1 should not displace ATG5 from RILP in an *in vitro* competition experiment. This experiment would confirm that a specific RILP binding to LIC1 Helix 1 impedes ATG5 binding.

Dynein Binding Site within RILP N-terminus

Interestingly, the Gassmann lab has also recently identified in their ongoing studies a dynein LIC1 binding site within adaptor proteins, including RILP. This site appears to be in the extreme N-terminal region of RILP, within the RILP homology (RH) domain (*personal communication*). The identification of dynein binding site within the RILP N-terminus provides an elegant tool to further map the minimal ATG5-binding region within the N-terminal domain of RILP. Given the premise of a direct LIC1 competition with ATG5, an addition of LIC1 Helix 1 should displace ATG5 from RILP N-terminus as well as RILP DBD (dynein binding domain), which would strongly suggest that the DBD is the binding site for ATG5. This experiment will also validate the minimal interaction region within RILP required for LIC1 binding. A further experiment with DBD and N-terminal RILP binding to recombinant purified ATG5 *in vitro* may show that ATG5 can bind equally efficiently to just the RILP DBD, as to the entire N-terminus, which would suggest that ATG5 binding site is located exactly within the dynein-binding site in RILP. It is also probable that RILP DBD may not show a decrease in ATG5 binding, upon addition of LIC1 Helix 1, which would show that ATG5 does not bind within the exact LIC1 binding site, and the competition arises not through direct displacement of ATG5 by LIC1 Helix 1 within a common RILP region, but rather through steric hindrance. In which case, RILP DBD may show less binding affinity for ATG5, than the entire RILP N-terminus *in vitro*. Further RILP N-terminal truncations to find the minimal common ATG5-dynein binding domain in RILP would then be necessary for mapping the minimal site. This set of experiments should identify the minimal binding region within RILP that coordinates its binding to two successive interactors ATG5 and LIC1 and further determine the nature of ATG5-dynein competition for RILP binding.

Cellular Regulation of RILP Binding to ATG5 vs Dynein

An interesting question that arises from competitive binding of ATG5 and dynein to RILP N-terminus is how RILP's interaction with either of these binding partners regulated? Which factors determine whether RILP should interact with LIC1 or ATG5? Our recent work shows that RILP is specifically upregulated and recruited to LC3-positive autophagosomes upon mTOR inhibition. By further extension of this observation, it is possible that changes in mTOR activity might alter not only RILP expression or behavior, but may also change its relative affinity for different interactors. For example, in steady-state cells, where autophagy functions at a basal level, RILP may predominantly interact with dynein and Rab7-GTP through its N and C- termini respectively. Thus, in homeostatic cells, RILP's function might be restricted to Rab7-positive late endosomal/lysosomal transport. However, upon mTOR inhibition and autophagy induction due to nutrient starvation, stress or other microenvironmental stimuli, RILP might show higher propensity to associate with the newly forming autophagosomes through an N-terminal ATG5 interaction and later, C-terminal LC3 interaction. There are two possible mechanisms for this switch in preferred interactors:

Hypothesis 1: Stochastic Increase in Available Membrane-bound ATG5 and Lipidated LC3 Promotes RILP Binding

A titration experiment with increasing ATG5 molar concentrations added to RILP DBD and LIC1 Helix 1 should show a progressively increased displacement of LIC1 and a corresponding increase in ATG5 binding. However, the relative binding affinities of LIC1 and ATG5 may differ, with LIC1 having a much higher affinity to RILP N-terminus. This experiment would show whether higher abundance of ATG5 triggers a switch in RILP interactors in cells. To determine whether it is specifically the membrane-bound ATG5 that controls the switch in RILP binding, ATG5-conjugated liposomes (in collaboration with T. Malia lab) can be used to perform the experiment.

Hypothesis 2: mTOR Kinase Phosphorylation of RILP N-terminus may Alter its Binding Affinity to ATG5 vs Dynein

mTOR kinase has been shown to negatively regulate autophagy by phosphorylating key early autophagy proteins like ATG101 and FIP200, and preventing their recruitment to AP initiation sites. It is possible that mTOR employs a similar, phosphorylation- dependent mechanism to regulate RILP N-terminal interaction with ATG5 or dynein at a given time. In this case, in steady state cells, an active mTOR kinase may phosphorylate RILP N-terminus, and prevent its binding to ATG5. Then dynein LIC1 will be preferentially able to bind RILP N-terminus and mediate retrograde transport. However, upon mTOR inhibition, RILP N-terminus would be dephosphorylated, and display an increased affinity for ATG5 and compete away LIC1. Within the extreme N-terminal domain of RILP, close to the putative DBD (RH1 domain), exists a predicted mTOR kinase phosphorylation site at Serine 15, within a consensus sequence of PGVPGWG**S**REAAGSA (source: PhosphoNET Kinase predictor). A mass spectrometric analysis for RILP N-terminal phosphorylated residues in control *vs* Torin1-treated neurons and non-neuronal cells would reveal whether S15 or other sites are differentially phosphorylated in mTOR inhibited cells. If so, expression of phosphomimetic (glutamic acid) RILP in cells would mimic an mTOR-activated state, and exhibit an increased binding affinity for LIC1, and a loss in ATG5 binding. Conversely, expression of the phosphomutant (alanine) RILP would display an increased binding to ATG5 and a corresponding loss in LIC1 binding, mimicking an mTOR-inhibited cell state. The relative binding affinities of phosphomutant and mimetic RILP can be directly tested with *in vitro* binding assays when added in equimolar concentrations. Again, we expect that the phosphomutant will display higher binding to ATG5, while the phosphomimetic will bind more potently to LIC1 Helix 1. Altogether, this line of analysis will help determine whether mTOR kinase regulates RILP interactions with its binding partners.

Conclusions

The set of experiments discussed in this project will help identify the minimal binding site in RILP for ATG5, dissect the mechanism for ATG5-dynein competition at RILP, and determine whether mTOR kinase regulates the relative binding affinities of ATG5 and dynein for RILP. Altogether, this line of analysis will identify molecular mechanisms underlying ATG5-Dynein competition for RILP.

4.2. Determining RILP Expression, Behavior and Function in Pathological Models with mTOR Dysfunction.

Our cellular data show that mTOR inhibition using short-term Torin1 treatment results in an upregulation of RILP expression as detected by mRNA levels, as well as a corresponding increase in the RILP protein levels. In cells, it is accompanied by a striking cytoplasmic redistribution of RILP onto the membranes of newly forming autophagosomes. These finding show that RILP expression and cellular behavior is responsive to mTOR kinase activity.

Our results suggest that modulation of mTOR kinase function in different cellular contexts would result in alteration of RILP function. Several neurodevelopmental and neurodegenerative disorders have been shown to display mTOR dysfunction. For example, in *Tuberous Sclerosis Complex* disease, mutations in TSC genes - negative regulators of mTOR - result in mTOR hyperactivation (Curatolo & Moavero, 2012). In addition, several other neurodevelopmental disorders such as *Autism Spectrum Disorder (ASD)*, Neurofibromatosis and Fragile X syndrome also show a hyperactivation of mTOR signaling cascade (Sato, 2016). mTOR kinase hyperactivation in these neurodevelopmental disorders might suppress the autophagy pathway in cells of the brain. Interestingly, our data now show that RILP expression and behavior is altered in response to mTOR activity. Thus, I hypothesize that mTOR kinase hyperactivation during neurodevelopment may alter RILP activity in these diseases. In TSC patient neurons, mTOR overactivation might result in decreased RILP expression or suppression of RILP recruitment to vesicular membranes. This would in turn suppress RILP-mediated autophagosome formation and transport, and result in cytoplasmic accumulation of p62- or Ubiquitin- labeled cellular cargoes marked for degradation through the autophagy pathway. This would lead to defects in neuronal homeostasis and lead to abnormal neuronal function. I expect that mTOR kinase inhibition in these neurons by pharmacological agents such as rapamycin or torin1 might restore RILP activity and in turn upregulate autophagy to induce clearance of autophagic cargo and alleviate disease pathology. Indeed, we already find RILP protein levels to increase in postnatal brain striatal tissue of systemically rapamycin treated mice, which is consistent with our cellular studies. Thus, mTOR inhibition *in vivo* induces RILP upregulation in the striatum, the brain region affected in ASD patients. To determine the effects of RILP upregulation, RILP intracellular localization in the striatal tissue and dissociated neurons of mock *vs* rapamycin-treated mice using an antibody to endogenous RILP in immunohistochemistry (IHC) would be useful. Upon Rapamycin treatment,

RILP should show punctate vesicular distribution in striatal neurons, particularly on ATG5-positive isolation membranes (phagophores) and newly formed LC3-positive autophagosomes. Additionally, a systemic mTOR inhibition in adult mouse striatal and cortical tissue using Torin1 injection might more robustly induce autophagy in brain tissue (Ori Leiberman, Dave Sulzer 2019), resulting in an even more striking RILP upregulation and redistribution in neurons. This treatment regimen could then be applied to a disease model with known mTOR dysfunction (TSC$^{+/-}$ mouse model). RILP mRNA and protein levels in TSC mice treated with Torin1 *vs* untreated controls could be monitored using immunoblotting and immunocytochemistry for RILP, ATG5, LC3 and Rab7. This experiment would determine whether RILP colocalizes with ATG5-positive isolation membranes, LC3-positive autophagosomes and Rab7-positive late endosomes in Torin1 –treated TSC mouse brains. mTOR inhibition might result in RILP-mediated restoration of autophagic turnover in TSC models.

Based on these results, it would be important to test whether ectopic RILP expression, which mimics mTOR inhibition, is sufficient for inducing autophagic activity in these disease models. This approach has great therapeutic value, since mTOR kinase has diverse roles including as a tumor suppressor, which restricts its amenability as a specific therapeutic target for autophagy. A downstream protein responsive to mTOR kinase which can selectively control autophagic-lysosomal function provides a better target for selectively treating autophagic dysregulation in the nervous system. For this purpose, RILP serves as an ideal candidate, since it can bind to ATG5, LC3, Rab7 as well as dynein, and associates with several intermediates in the autophagy pathway including the isolation membrane, the nascent and mature APs, and late endosomes. To test whether ectopic RILP expression is sufficient for offsetting the effects of mTOR overactivation in suppressing autophagy pathway, fluorescently tagged RILP cDNA could be lentivirally delivered to TSC animals perinatally and the tissue probed for levels of autophagy by measuring LC3-I/II ratios and by counting number of APs per cell. An increase in RILP-mediated autophagy in these tissues, should overcome the deficits caused by mTOR hyperactivation in these models, and might alleviate pathological symptoms observed in TSC brains.

Conclusions

Altogether, these experiments will determine the effects of pharmacological inhibition of mTOR kinase on RILP behavior *in vivo*. They will help determine whether mTOR controls RILP function and whether RILP upregulation can induce autophagy *in vivo*, which can then be adapted as a therapeutic regimen for treating neurodevelopmental diseases with known mTOR dysfunction.

4.3. RILP Roles in Dendritogenesis, Remodeling and Spine Pruning.

Neuronal communication in the cerebral cortex depends on pre-synaptic axonal terminals contacting the post-synaptic neuronal dendrites to form excitatory synapses. Synapses are formed and remodeled throughout an organism's lifespan to incorporate new information, learning and memory. On the post-synaptic side, specialized microstructures called dendritic spines - 1-5 micron long protrusions that arise from the dendritic shaft and constitute an electrochemically and spatially separated compartment - serve as the principle site for receiving input from pre-synaptic neurons. Dendritic spines are highly dynamic structures that are formed and eliminated throughout the lifespan of an organism. The composition, morphology and electrochemical properties of dendritic spines have been under investigation for decades, however the basic cellular and molecular mechanisms underlying spine formation, function and turnover remain surprisingly poorly understood. Emerging evidence now suggests that autophagy and endo-lysosomal pathways - two principle mechanisms for recycling and degrading cellular materials- might also control synaptic remodeling (Kallergi et al., 2020; Vassiliki Nikoletopoulou, Kyriaki Sidiropoulou, Emmanouela Kallergi, Yannis Dalezios, & Nektarios Tavernarakis, 2017).

Defects in dendritic spine formation or remodeling have been implicated in several neurodevelopmental and psychiatric disorders including autism spectrum disorders (ASD), bipolar disorder (BP) and schizophrenia, among others. Schizophrenia, a complex neuropsychiatric disease, affects 1% of the world's population and is characterized by deficits in motivation, perception and cognition (Ross, Margolis, Reading, Pletnikov, & Coyle, 2006). Clinical analyses have revealed decreased brain volume and mislocalization of neurons, especially in the prefrontal cortex, reduced complexity of dendritic arbor and decreased dendritic spine density which altogether classify schizophrenia as a subtle neurodevelopmental disorder(Kubicki et al., 2005; Spencer et al., 2003). Although mutations in several genes have been identified in schizophrenia, the molecular and cellular mechanisms that go awry in this disease have remained mostly unknown. Autism spectrum disorders affect about 2% of all children in the United States, and are also characterized by abnormalities in dendritic spine density and behavior. Unlike Schizophrenia, ASD brains often show an increased spine density, and excess synapse formation, particularly in the layer III neurons of the frontotemporal cerebral cortex. Seminal work from Dr. David Sulzer's group at Columbia has shown that mTOR overactivation in TSC models of autism spectrum disorder results in decreased autophagy in neurons, and excess spine density. This phenomenon can be reversed by rapamycin-mediated mTOR inhibition that upregulates autophagy - as measured through LC3-II/I ratios - and restores normal spine density in the cerebral cortex. More importantly, rapamycin treatment of these TSC mice also shows correction in the autistic behavior exhibited by these animals. This work provides compelling evidence that mTOR kinase might regulate autophagic dendritic spine pruning in cortical neurons, however the molecular machinery involved in this process remains to be identified (G. Tang et al., 2014). These data, combined with our observations, suggest that RILP might form a mechanistic link between mTOR signaling and autophagy pathway in dendrites. What is the molecular mechanism of dendritic spine pruning regulated by mTOR kinase? Does RILP contribute to this process by retrieving materials from spines *via* autophagic and endosomal pathways in dendrites?

To Determine the Molecular Mechanism of RILP-Mediated Autophagic Spine Pruning in Dendrites

Newly emerging evidence shows autophagy pathway to be critical for selective degradation of scaffolding proteins, as well as neurotransmitter receptors at the synapse (Kallergi et al., 2020; Vassiliki Nikoletopoulou et al., 2017; Nikoletopoulou & Tavernarakis, 2018). This highly regulated autophagic turnover of pre- and post-synaptic proteins is thought to play an important role in synaptic maintenance and for long-term depression (LTD)-mediated spine elimination. Conversely, Long-term potentiation (LTP) suppresses autophagy, resulting in spine stabilization and maturation (Vassiliki Nikoletopoulou et al., 2017). Although autophagy seems critical for controlling both pre- and post-synaptic composition, the basic mechanisms, cytoskeletal motors and effector proteins involved in this process are only beginning to be understood. Our preliminary studies show extensive RILP punctate distribution in axons as well as dendrites- including some PSD-95 loci - in mature DIV22 cortical neurons. How RILP controls autophagy at synapses, and whether this is a steady-state or stress-induced function for RILP remains to be tested.

To determine whether RILP-mediated autophagy is essential for dendritic spine maintenance and function, mature DIV22 rat cortical neurons could be immunostained for endogenous RILP, LC3 (APs), Rab7 (LEs) and ATG5 (phagophores, i.e. incomplete APs) distribution at pre- and postsynaptic densities visualized by Bassoon and PSD-95 respectively in MAP2-positive dendrites. Lentiviral RILP silencing would show whether RILP is required for normal dendritic arborization and spine or synapse formation/ maturation. Spine density, spine morphology- filopodia-like/ stubby/ mushroom/ long; and PSD-95, AMPA/ NMDA receptors, and Kalirin-7 distribution in dendrites (Arellano, Benavides-Piccione, DeFelipe, & Yuste, 2007; Heck & Benavides-Piccione, 2015) could be tested. If RILP is critical for dendritic spine pruning, then its depletion would result in excess spine density, an abnormal accumulation of PSD proteins turned over by autophagy and alterations in spine morphology and Kalirin-7 distribution. To further confirm if it is RILP's role in autophagy that is specifically required for pruning, RILP RNAi rescue with wild type (wt) *vs* LIR3- mutant RILP could be performed. The wt but not LIR-mutant RILP should be able to restore spine density. Lentiviral overexpression of RILP in cultured *wt* rat neurons may increases spine elimination. Finally, vGlut- pHluorin live imaging in wt *vs* RILP-depleted neurons using action potentials evoked by passing 1millisec bipolar current would show whether RILP is required for normal synaptic function. Alternatively, NMDAR and mGluR1-mediated LTD (Kallergi et al., 2020) using a short pulse of NMDA and DHPG, respectively; or BDNF-induced LTP (Vassiliki Nikoletopoulou et al., 2017) in neurons would show whether dendritic RILP, LC3, ATG5 and Rab7 distribution is altered upon these stimuli. Induction of LTP/LTD would be confirmed by AMPA receptor subunit GluA2 (GRIA2) distribution on cell surface. An increase in RILP puncta in dendritic shaft and spines might be detected upon LTD but not LTP, which would be blocked upon RILP silencing. Finally, two-photon Ca^{+2} uncaging in GFP-RILP and RFP-LC3 expressing hippocampal or cortical live rodent brain slices, may reveal whether RILP-positive APs are specifically recruited to spines undergoing LTD, but are excluded from those undergoing LTP within the same dendrite.

Altogether, these experiments will elucidate RILP function in dendritic autophagy, providing critical proof-of-principle data for targeting RILP activity *in vivo* during neurodevelopment.

To Test Whether mTOR Controls RILP Behavior in Spine Pruning During Neurodevelopment

mTOR kinase is a key negative regulator of autophagy. Upon environmental stress, nutrient starvation or acute injury, mTOR is inhibited which upregulates autophagy in non-neuronal cells (Robert A. Saxton & Sabatini, 2017). My recent work shows that mTOR inhibition upregulates RILP expression and recruits it to newly forming autophagosomes in DIV14 cortical neurons. mTOR is overactivated in autism spectrum disorders (ASD), and has been shown to suppress autophagy leading to decreased spine pruning (Guomei Tang et al., 2014). Inhibition of mTOR in an ASD mouse model ($Tsc2^{+/-}$) restores normal spine density and corrects autism-like behavioral patterns in these animals, (work of collaborator Dr. David Sulzer), however the autophagic machinery involved in this process remains to be identified (Guomei Tang et al., 2014). Given the basic mTOR-controlled role for RILP in neuronal autophagy that I have found, it will be of interest to test whether RILP contributes to mTOR-mediated spine pruning in ASD. We find that systemic injection of rapamycin in Tsc2$^{+/-}$ mice upregulates RILP protein in mouse brain tissue, suggesting that mTOR overactivation in ASD might downregulate RILP function and, in turn, suppress RILP-mediated autophagy leading to pruning deficits. Thus, genetic or pharmacological upregulation of RILP may activate autophagy to help restore spine pruning in Autism Spectrum Disorders.

Whether mTOR inhibition induces RILP-mediated spine pruning in dendrites can be tested in wt DIV22 rat cortical neurons treated to a short 6-8 hour Torin1 or Rapamycin treatment. mTOR-induced changes in RILP expression/ distribution in dendrites could be monitored using high-resolution spinning disk confocal live imaging and immunostaining. Live imaging might reveal whether spines enriched for RILP get pruned over time. A systemic (intraperitoneal) administration of rapamycin to Tsc2$^{+/-}$ mice for seven days (P21→P28), followed by immunostaining for RILP in dendrites would help quantify spine density and relative RILP colocalization with ATG5, LC3 and Rab7 in control *vs* mTOR-inhibited neurons. AAV-mediated GFP-RILP delivery to Tsc2$^{+/-}$ mice *via* early postnatal ventricular injection and quantification of the spine density in RILP-expressing *vs* control brains at P28 i.e. after first wave of developmental pruning in the striatum, would determine if RILP upregulation alleviates pruning deficits in ASD mouse brains. A comparison of relative numbers of LC3 and ATG5 puncta in AAV-RILP *vs* untreated animals, and the ratios of LC3-I/II on western blot would determine whether RILP overexpression upregulates autophagy in ASD brains *in vivo*. Finally, the exact contribution of RILP to pruning can be determined using a LIR-mutant *vs* Rab7-mutant RILP delivery to Tsc2$^{+/-}$ mice. Based on the outcomes of these experiments, animal behavior studies could be performed to test for display of social novelty, social preference and repetitive behavior.

Altogether, these experiments will provide valuable insights into RILP function *in vivo* upon mTOR inhibition and validate the therapeutic potential of RILP in neurodevelopmental diseases with known dendritic spine alterations.

4.4. RILP Roles in Clearance of Neuropathological Aggregates.

My recent work shows that RILP-mediated autophagy is required for p62/Sequestosome-1 turnover in cells. Cytoplasmic p62 aggregates are a key feature of neurodegenerative pathology in Alzheimer's, Huntington's and Amyotrophic Lateral Sclerosis. Molecular basis for p62 aggregation in these diseases is not well understood and has been attributed to defects in AP/LE function (Liu & Li, 2019), or to changes in p62 expression (Caccamo et al., 2017a), although the exact molecular mechanisms that go awry remain to be identified. It is possible that RILP upregulation in neurodegeneration may enhance aggregate clearance and restore AP-LE function.

In Alzheimer's Disease brain tissue, an accumulation of immature autophagic vesicles has been reported in dystrophic neurites in the brain before the appearance of NFTs and amyloid beta plaques (R. A. Nixon et al., 2005; W. H. Yu et al., 2005). The widely-studied AD mouse models (APP/PS1 and J20) each show p62 aggregation (Kuusisto et al., 2002). How autophagic dysfunction results in onset of AD pathology, however, remains poorly understood. Since RILP contributes to protein homeostasis, it might well play a role in protein clearance and serve as a novel molecular tool to modulate maturation and turnover of AP/LEs in AD brains. In preliminary experiments, we have found RILP expression to be high during early mouse brain development, but decrease ~4-fold in adulthood. In collaboration with Dr. F. Bartolini at Columbia University, we have found RILP levels to be strikingly reduced in moderate and severe human AD brain tissue compared to age-matched controls. This reduction in RILP might well suppress AP/LE function and turnover, inducing AD pathology. Interestingly, RILP levels are upregulated in severe AD pathology suggesting that protein aggregation might upregulate RILP expression however, at later stages of this disease elevated RILP levels may not be effective for aggregate clearance. These observations suggest that early autophagic upregulation might aid in aggregate clearance and help prevent severe pathology. To further test this hypothesis, familial Alzheimer's disease (FAD) knock-in ES cell line-derived DIV60 cortical neurons can be used study endogenous RILP behavior in these neurons.

Live imaging of lentivirally infected GFP-tagged RILP and RFP-tagged LC3 or Rab7 in FAD neurons, would reveal early defects in AP/LE formation, transport or turnover, if any. Immunostaining for endogenous RILP, ATG5, LC3, and Rab7 would show relative co-distribution and colocalization of RILP with vesicular compartments in control *vs* disease neurons. It will also be important to test the effects of lentivirally delivered wt *vs* RILP LIR- and Rab7- mutants on amyloid beta and p62 turnover in FAD neurons in real time. RILP protein distribution can be analyzed in J20 or APP/PS1 mouse brain slices (in collaboration with Dr. F. Bartolini) at early, moderate and late stages of disease progression, to test whether RILP association with ATG5 (phagophore), LC3 (APs) or Rab7 (LEs) is disrupted at different stages of the disease. It will be useful to test whether RILP associates with amyloid beta aggregates or neurofibrillary tangles in cultured neurons using immunofluorescence and AD mouse model brain tissue using immunohistochemistry. Based on the outcomes of these experiments, AAV-assisted RILP delivery to J20 mice could be used to test whether RILP upregulation delays or reduces amyloid beta plaque and NFT formation. A time course of RILP delivery to adult animals at different ages could also be adopted to determine whether a particular time window is therapeutically beneficial in reversing amyloid beta pathology and optimizing neuronal survival.

Increasing evidence suggests that autophagic dysfunction in AD might result from defects in neuronal lysosome activity. As discussed in previous sections, RILP associates with Rab7-positive late endosomal structures as well as with the lysosomal proton pump V-ATPase subunit V1G1. RILP is required for the recruitment and stabilization of V1G1 subunit on late endosomal and lysosomal membranes. Depletion of RILP prevents assembly of the vacuolar V-ATPase, and its recruitment to lysosomal compartments, which prevents lysosomal acidification and decreases its function. While there is some evidence for lysosomal dysfunction in AD patient neurons, the exact molecular pathology remains poorly understood. For example, the function of Cathepsin enzymes and V-ATPase on LAMP1-positive lysosomes in AD neurons remains to be carefully examined. It is possible that RILP downregulation in AD neurons results in V1G1 mislocalization to golgi compartments and a failure in V-ATPase assembly on lysosomal membranes. This would result in low acidification of lysosomes and accumulation of degradative cargoes in AD neurons. In such a scenario, ectopic expression of RILP might help restore V1G1 assembly into functional V-ATPase and its delivery to LAMP1-positive lysosomal membranes. RILP activity in AD neurons could be elevated using mTOR inhibitors such as rapamycin or torin1, which are already FDA-approved for treatment of cancers. Interestingly, another study of human early, moderate and late stage AD patient brains has shown that mTOR is hyperphosphorylated (pS2448) and autophagy is suppressed during early stages of the disease (Tramutola et al., 2015). This again suggests that RILP activity might be suppressed at early stages of the disease, which is reflected in our western blot analysis of AD patient brain tissue at different stages of disease progression (Figure 28). Thus, restoration of RILP activity might help overcome defects in lysosomal functioning in AD neurons and prevent cytoplasmic aggregation of autophagic compartments and amyloid beta. It will be important to test the expression of Cathepsin proteases, V-ATPase subunits, LAMP1 and LAMP2 in AD brain tissue during disease progression using western blot and subcellular distribution of V-ATPase with LAMP1 vesicles using immunostaining. This would determine whether the proton pump is getting properly assembled onto the lysosomal membranes. If the expression of V-ATPase is reduced in AD, the V-ATPase subunits could be ectopically expressed in AD neurons using lentiviral delivery. If the V-ATPase is mislocalized within AD neurons, RILP expression might promote V-ATPase delivery to lysosomal membranes through dynein-mediated transport of V-ATPase from golgi to lysosomal membranes. If RILP upregulation can increase autophagy-lysosomal functioning in AD neurons, it will present a very attractive therapeutic target for alleviating AD pathology. Altogether the above experiments will determine whether targeted RILP upregulation has a therapeutic potential for AD and other aggregate-prone degenerative diseases.

4.5. Characterizing the Role of Dynein Regulatory Proteins LIS1 and NudE/EL in RILP Behavior and Function *In Vivo*

Our cellular and biochemical data shown in *Chapter 3* strongly suggest that LIS1 is an obligate component of the RILP-dynein-dynactin supercomplex. Based on our preliminary results, we hypothesize that depletion of LIS1 might affect RILP behavior in neuronal and non-neuronal cells, and adversely affect RILP-mediated autophagy in these cells. Whether LIS1 participates in autophagosome transport is completely unknown, however, there is a possibility that neurodevelopmental defects seen in patients with LIS1 mutations might also arise due to its roles in the autophagy pathway.

This hypothesis could be tested using LIS1 silencing in cortical neurons and testing GFP-RILP behavior at several different stages of *in vitro* neuronal development, for example, at DIV 3, 10 and 21. Imaging at different time-points during *in vitro* differentiation would provide unique insights into RILP function over the course of neuronal differentiation. For example, it has been shown that at low DIVs in culture, axonal lysosomal transport is mostly retrograde; however, upon maturation, it becomes increasingly bidirectional, but with bias towards retrograde (unpublished data in our lab). During differentiation, the axonal architecture develops and the microtubule cytoskeleton undergoes posttranslational modifications, factors that can alter motor binding affinities to microtubules and their processivity within axons. Furthermore, at DIV21, the dendritic compartment is well-defined and shows formation of synapses; which would allow for a comparative analysis of RILP distribution, directionality, motility and cargo specificity within dendrites *vs* axons. If indeed LIS1 is required for RILP-mediated motility, stalling, slower velocity or decreased travel distances for RILP vesicles would be observed. Previous work from our own lab has shown that LIS1 depletion in neurons resulted in a stalling of only the larger vesicles, but did not alter the motility of smaller vesicles, suggesting a LIS1 role in larger, heavy-load vesicular cargo transport (Yi et al., 2011). Based on these results, it is possible that within the RILP-positive vesicle population, there would be a size-specific effect of LIS1 depletion, exclusively on the larger, heavy-load subset of RILP vesicles. A question arises as to whether RILP selectively recruits LIS1 as means for 'force adaptation' at the larger vesicles or whether it recruits LIS1 as an obligate factor in the dynein supercomplex, irrespective of the vesicular size.

It would also be important to test RILP behavior in NudE/EL depleted neurons. A similar set of experiments as discussed above would determine the effect of NudE/EL shRNA-mediated depletion on run lengths, directionality and velocity of RILP dual-positive vesicles. Based on our cellular data, we predict that NudE/EL depletion will not affect RILP behavior in neurons, which would suggest that NudE/EL are dispensable for RILP-mediated vesicular transport. An analysis of directionality and travel distances of autophagosomes (LC3$^+$), early endosomes (EEA1$^+$), late endosomes (Rab7$^+$) and lysosomes (LAMP1$^+$) in NudE/EL depleted neurons would reveal whether NudE/EL silencing affects motility of diverse neuronal vesicular cargoes.

4.6. Determining the Contribution of LIS1 to Different Cellular Cargoes of RILP

We have shown that RILP associates with both, Rab7-positive late endosomes as well as LC3-positive autophagosomes in neuronal and non-neuronal cells. Autophagosomes are much larger in size compared to late endosomes and lysosomes (Yim & Mizushima, 2020), and RILP exhibits the ability to associate with both, larger heavy-load cargoes like autophagosomes (LC3$^+$) and the smaller low-load cargoes like late endosomes (Rab7$^+$). Whether the composition of dynein supercomplex recruited by RILP to LEs and APs is the same, remains to be tested. This could be tested by determining whether LIS1 depletion affects retrograde transport of both RILP-LC3$^+$ APs and RILP-Rab7$^+$ LEs in neurons. A sucrose gradient centrifugation for isolating autophagosomal and endosomal membrane fractions to determine the relative amounts of RILP, LIS1, NudE/EL and dynein subunits would shed light on whether the composition of RILP motor complex in cells differs for the higher-load early APs *vs.* the lower-load LEs. Recent studies have shown that Individual LIS1- and Nde1/Ndel1-decorated lipid droplets can be detected *in situ* by immunocytochemistry (Reddy et al., 2016). A similar *in vitro* approach could be adopted to test for the presence of dynein regulatory proteins on Syntaxin17-positive APs *vs* SNAP29- positive LEs. Fluorescently-tagged LIS1 and RILP co-behavior in cortical neurons along with either Rab7 or LC3 would determine, using three-channel live imaging, whether RILP and LIS1 colocalize and comigrate with APs and/or LEs in axons. As additional markers for distinguishing between APs and LEs, AP- and LE- specific SNARE proteins, Syntaxin17 and SNAP29, could be used. If these experiments show that LIS1 is indeed required for RILP-mediated AP/LE transport in neurons, effects of LIS1 depletion or patient mutations (Saillour et al., 2009) on autophagic cargo clearance in neurons could be tested using p62/ Sequestosome-1 and ubiquitinated cargo turnover.

Altogether, these experiments will determine whether LIS1 is an obligate dynein regulator required for RILP-mediated vesicular transport in neuronal and non-neuronal, and whether LIS1 recruitment by RILP is cargo-specific. Outcomes from these experiments will also evaluate the roles for LIS1 in neuronal autophagic turnover, a novel function that may have pathogenic potential.

4.7. Characterizing the Role of LIS1 in RILP-Dynein Supercomplex Behavior *In Vitro*

To determine the mechanochemical properties of the RILP supercomplex at single-molecule resolution, we have engineered a SNAP-eGFP-2xStrep-RILP construct for bacterial expression and affinity purification. This construct would be used to purify recombinant RILP, by recruiting it to streptavidin beads and then to pull down dynein, dynactin, and LIS1 from adult rat brain lysates. The composition of the complex could also be controlled by assembling it *in vitro* using the recombinant RILP, baculovirus-expressed LIS1 purified in our lab (McKenney et al., 2010), and tandem-tagged affinity purified dynein and dynactin complexes expressed In Hela cells (Splinter et al., 2012). Run lengths, directionality and velocity of the RILP complex could be determined *in vitro*. Force measurements for individual RILP complexes could be generated using a laser trap set-up. The composition of individual RILP complexes could be determined using size-exclusion chromatography, fluorescence intensity measurements and photo-bleaching assays, for e.g. one-step photobleaching indicates a single RILP complex. LIS1 contribution to RILP function could be tested by preparing RILP complexes from either *wt vs* LIS1 immuno-depleted cytosol, or using *in vitro* assembled RILP complexes prepared with or without recombinant LIS1.

Altogether this cellular and single molecule analysis will determine the specific contribution of LIS1 to RILP- mediated retrograde transport *in vivo* as well as the biophysical properties of this complex *in vitro*.

CHAPTER 5. CONCLUDING REMARKS

This work elucidates a novel, mTOR-sensitive neuronal and non-neuronal autophagy pathway controlled by the dynein adaptor protein, RILP. This basic pathway has broad implications for several fundamental cell biological processes, including dendritogenesis and dendritic dynamics in neurodevelopment, ubiquitinated cargo turnover during degeneration as well as heavy-load cargo transport in cells. Modulation of RILP-mediated autophagy pathway might be key to treating neurodevelopmental and degenerative disorders with mTOR dysfunction and cytotoxic protein aggregation. Future directions outlined in this study on RILP expression and function in physiological and disease conditions will provide valuable insights into therapeutic potential of RILP in regulating autophagy *in vivo*.

Finally, this work also sets a precedent for other dynein adaptor proteins. This class of proteins may not only control dynein motor behavior in cells, but might also contribute to the coordination and regulation of complex biological pathways to maintain cellular homeostasis and promote survival.

CHAPTER 6. MATERIALS AND METHODS

EXPERIMENTAL MODEL AND SUBJECT DETAILS

Rat Cortical Neuron Culture and Transfection

All experiments involving animals were performed in accordance with protocols approved by the Columbia University Medical Center Institutional Animal Care and Use Committee. E19 embryonic brains from pregnant Sprague-Dawley female rats were used to prepare cortical neuronal cultures as described previously (Yi J., Khobrekar N. et.al. 2016). Briefly, embryonic brains were dissected in ice-cold HBSS (ThermoFisher Scientific #24020117) supplemented with 10mM HEPES (pH 7.4), 10mM $MgCl_2$, 1% P/S and 0.5mM L-glutamine. The hippocampal tissue was discarded and only the cortical tissue was trypsinized for 15 min at 37°C in water bath and resuspended in DMEM with 10% FBS and 1% P/S. The dissociated neurons were spun down to remove debris, counted and plated onto MatTek 50mm glass bottom dishes, pre-coated with Poly-L-Lysine (PLL) (1mg/mL) and Laminin, at a density of 180,000 for live imaging experiments. ~12-15 hr post-plating, the medium was changed to neurobasal (Gibco) supplemented with 0.5mM L-Glutamine, B-27 serum free supplement (Gibco) and 1% P/S. The media were changed every 2-3 days. During Torin1 treatment, the medium was changed to neurobasal with insulin-free B27 supplement (Gibco). For immunofluorescence experiments, neurons were plated on 18x18mm coverslips or 35mm Glass bottom dishes coated with PLL and Laminin.

For live imaging experiments, DIV7-8 cortical neurons were transfected using magnetofectamine kit (OZ biosciences # MTX0750). 0.8μg of total DNA, 5μL of Lipofectamine 2000 and 0.75μL magnetic bead slurry were used to transfect one 50mm glass bottom dish of neurons, half of the medium was changed after 15 minutes and neurons were imaged after 15 hr. Cells with low expression of tagged proteins were selected for motility analysis. 2μg of RILP shRNA (rat RILP shRNA: 5'-GGAGGCTTAACTCTGGGTTCC- 3') was used to transfect $1x10^6$

cortical neurons with nucleofector kit (Lonza) and an amaxa nucleofector II (program O-003). Successfully transfected neurons were imaged 72 hr after knockdown at DIV3.

Cell Culture

C6 rat glioma cells (ATCC # CCL-107) were cultured in DMEM supplemented with 10% FBS and 1% penicillin/streptomycin (P/S), at 37°C with 5% CO2. Transfection of C6 cells was performed using a nucleofector kit V (Lonza) with an Amaxa Nucleofector II set to program U-030 or U-009, according to the manufacturer's instructions. Cells for qRT-PCR analysis were collected 72 hr following transfection and sorted using fluorescence-activated cell sorting (FACS) to isolate GFP/RFP-positive cells.

METHOD DETAILS

Drug Treatments

mTOR inhibitor Torin1 (Tocris Biosciences) was resuspended in DMSO and used at final concentrations of 0.5 µM for short term 1-2 Hr treatments in C6 cells or rat cortical neurons (DIV13) for 6 Hr at 37°C. Bafilomycin A1 (Selleck Chemicals, #S1413) was resuspended in DMSO and used at final concentration of 100nm for 2-6 Hr in combination with Torin1 in C6 glioma cells.

Immunoprecipitation and Western Blots

For immunoprecipitation experiments, C6 cells were transfected with 5µg plasmid DNA, harvested between 15-17 hr post-transfection and pelleted at 1200 rpm for 10 min. Cell pellets were flash-frozen in liquid nitrogen and then lysed in IP buffer (50mM HEPES, 50mM PIPES, 2mM MgCl2, 1mM EDTA) supplemented with 1% NP-40, 1mM DTT, 1:100 protease inhibitor cocktail (Sigma-Aldrich #P8340) and 1:100 Phosphatase Inhibitors (ThermoFisher Scientific #78420) for 30 min on ice. Lysates were spun at 14,000 rpm at 4°C for 15 min and the supernatants were used for immunoprecipitations. For one immunoprecipitation, 30-40µL of GFP Trap-MA bead slurry (anti-GFP V_HH coupled to magnetic agarose beads, ChromoTek) was used for 1-1.25 mg of total protein lysates. The pre-clearing of the lysate was performed using 40µL protein A Dynabeads (ThermoFisher Scientific) for 1 Hr at 4°C with constant mixing. The GFP-trap beads were then incubated with pre-cleared cell lysates for 2 hr at 4°C, with constant mixing. The beads were washed three times with lysis buffer and boiled at 95°C for 3 min in 1X laemmli buffer supplemented with beta-mercaptoethanol (2%).

For Western blotting, cells were lysed on ice for 30 min in RIPA Buffer (50mM Tris-HCl; 100mM NaCl; 1mM EGTA; 1% NP-40, pH 7.4), supplemented with 1:100 protease inhibitor cocktail, 1mM DTT and 1% NP-40. The lysates were loaded onto Bis-Tris gradient (4-20%) SDS–polyacrylamide gels (Life Sciences), along with Precision Plus Protein Dual Color Standards (161-0374; Bio-Rad). Proteins were transferred to 0.2µM polyvinylidene difluoride (PVDF) membranes (Immobilon-P #ISEQ00010; Millipore) overnight, and blocked in 5% milk for 2 hr at room temperature. Membranes were probed with primary antibodies diluted in 1% milk at 4°C, overnight. After washing with PBS, membranes were incubated with fluorescently tagged secondary antibodies (Rockland and Life Sciences) and scanned in an Odyssey system (LICOR Biosciences) version 3.0. Protein levels were quantified using ImageJ (NIH, Bethesda, MD).

In Vitro Binding Assays

Vectors for GST-tagged Full length and N-terminal RILP fragment were transformed into BL21-Gold(DE3) pLysS *Escherichia coli* (Agilent Technologies) and expressed for 4 hours at 18°C after induction with 0.1mM IPTG. Purification was performed following standard procedures involving application of the bacterial lysates to MagneGST Glutathione particles for 3 hr at 4°C on rotation in binding buffer (300mM NaCl, 50mM Tris, 1mM DTT, Lysozyme 1 mg/ml, cOmplete protease inhibitor EDTA-free, Roche at 1 mg/mL, DNase 10µg/ mL and 0.1% NP-40, pH 7.4). Bound proteins were eluted in elution buffer (150mM NaCl, 50mM Tris and 1mM DTT, supplemented with 50mM reduced glutathione, pH 7.4). Overnight dialysis was performed in the elution buffer to remove glutathione from the eluates.

In vitro binding was performed using 0.5mM GST-tagged RILP constructs and recombinant human His-tagged ATG5 (Abcam # ab103787) in binding buffer (100mN NaCl,

50mM Tris, 1mM DTT and 1% NP-40) for 3 hr at 4°C. The proteins were blotted for using anti-GST and anti-ATG5 antibodies.

In vitro competition assay was performed using 1μm GST-tagged RILP, His-tagged ATG5 and Myc-tagged LIC1 in Tris binding buffer (pH 7.4). Magnetic Glutathione beads were used for pulldowns after 3 hr incubation at 4°C.

Immunofluorescence

Coverslips were washed twice with Tissue culture grade PBS to remove traces of culture medium. Cells were fixed in 4% Paraformaldehyde in PHEM buffer (120mM PIPES, 50mM HEPES, 20mM EGTA, 4mM MgSO4, pH = 7.4) for 12-15 min at room temperature (RT), followed by permeabilization with 0.3% Saponin in PHEM buffer for 10 min at RT. Cells were blocked in 1% BSA in PBS or PHEM for 1 hr, stained with primary antibodies O/N at 4°C, washed for 30 min in PBS and then stained with fluorophore-conjugated secondary antibodies (Jackson laboratories) for 1 hr at room temperature. The coverslips were mounted in Aqua Poly/Mount (PolyScience #18606) and allowed to cure overnight at room temperature before imaging. Images were acquired using an IX83 Andor Revolution XD Spinning Disk Confocal with 60x silicone oil objective (NA 1.30) and a 2x magnifier coupled to an iXon Ultra 888 EMCCD Camera. Final images were compiled using ImageJ (NIH, Bethesda, MD) or Fiji.

Quantitative RT-PCR analysis

Cell lysis, mRNA to cDNA synthesis and quantitative PCR were performed using the Power SYBR Green Cells-to-Ct Kit (Ambion/ThermoFisher). Reactions were performed in a QuantStudio 6 Flex Real- Time PCR System (ThermoFisher). Primers were designed to have TMs around 60°C and to generate amplicons of 70 to 200 base pairs, ideally separated by at least one intron. Three replicates were performed for each condition per experiment and the experiments were carried out at least in triplicates. Target cDNA levels were analyzed by the comparative cycle (Ct) method and values were normalized against β-ACTIN, HPRT and YWHAZ housekeeping gene expression levels.

Live Imaging and Motility Analysis

Live imaging was performed using an IX83 Andor Revolution XD Spinning Disk Confocal System with an environmental chamber at 37°C, a 60x oil objective (NA 1.30) and a 2x magnifier coupled to an iXon Ultra 888 EMCCD Camera. Images were taken at the rate of 1 frame/sec for 3 or 5 min. Kymographs were generated with ImageJ (NIH) to quantify the distance travelled, velocity and direction of transport of the individual vesicles. Individual runs were counted if the pause between two processive movements was less than 10 sec. If the pause was longer, then the runs were counted as separate.

Quantification and Statistical Analyses

For all experiments involving rat neurons, triplicates were obtained from three different pregnant rats. sample sizes are specified in figure legends. All statistical analyses were performed in Prism 6 (GraphPad Software, La Jolla, CA, USA). Normality tests were performed to determine use of parametric tests. For comparing two populations, two-tailed unpaired t-test with Welch's correction was used. For multiple comparisons, ANOVA with Tukey's multiple comparison test was used. $p < 0.05$, $p < 0.01$, $p < 0.001$ values are labeled with *, ** and ***, respectively. Error bars represent mean $\pm$ SEM unless stated otherwise.

REFERENCES

Abada, A., & Elazar, Z. (2014). Getting ready for building: Signaling and autophagosome biogenesis. *EMBO reports, 15*(8), 839-852. doi:10.15252/embr.201439076

Abe, N., Borson, S. H., Gambello, M. J., Wang, F., & Cavalli, V. (2010). Mammalian target of rapamycin (mtor) activation increases axonal growth capacity of injured peripheral nerves. *J Biol Chem, 285*(36), 28034-28043. doi:10.1074/jbc.M110.125336

Alemu, E. A., Lamark, T., Torgersen, K. M., Birgisdottir, A. B., Bowitz Larsen, K., Jain, A., . . . Johansen, T. (2012). Atg8 family proteins act as scaffolds for assembly of the ulk complex: Sequence requirements for lc3-interacting region (lir) motifs. *Journal of Biological Chemistry.* doi:10.1074/jbc.M112.378109

Ally, S., Larson, A. G., Barlan, K., Rice, S. E., & Gelfand, V. I. (2009). Opposite-polarity motors activate one another to trigger cargo transport in live cells. *J Cell Biol, 187*(7), 1071-1082.

Alvarez-Erviti, L., Rodriguez-Oroz, M. C., Cooper, J. M., Caballero, C., Ferrer, I., Obeso, J. A., & Schapira, A. H. (2010). Chaperone-mediated autophagy markers in parkinson disease brains. *Arch Neurol, 67*(12), 1464-1472. doi:10.1001/archneurol.2010.198

Arellano, J., Benavides-Piccione, R., DeFelipe, J., & Yuste, R. (2007). Ultrastructure of dendritic spines: Correlation between synaptic and spine morphologies. *Frontiers in Neuroscience, 1*(10). doi:10.3389/neuro.01.1.1.010.2007

Axe, E. L., Walker, S. A., Manifava, M., Chandra, P., Roderick, H. L., Habermann, A., . . . Ktistakis, N. T. (2008). Autophagosome formation from membrane compartments enriched in phosphatidylinositol 3-phosphate and dynamically connected to the endoplasmic reticulum. *J Cell Biol, 182*(4), 685-701. doi:10.1083/jcb.200803137

Axe , E. L., Walker , S. A., Manifava , M., Chandra , P., Roderick , H. L., Habermann , A., . . . Ktistakis , N. T. (2008). Autophagosome formation from membrane compartments enriched in phosphatidylinositol 3-phosphate and dynamically connected to the endoplasmic reticulum. *The Journal of Cell Biology, 182*(4), 685-701. doi:10.1083/jcb.200803137

Bar-Peled, L., Chantranupong, L., Cherniack, A. D., Chen, W. W., Ottina, K. A., Grabiner, B. C., . . . Sabatini, D. M. (2013). A tumor suppressor complex with gap activity for the rag gtpases that signal amino acid sufficiency to mtorc1. *Science, 340*(6136), 1100. doi:10.1126/science.1232044

Bar-Peled, L., Schweitzer, Lawrence D., Zoncu, R., & Sabatini, David M. (2012). Ragulator is a gef for the rag gtpases that signal amino acid levels to mtorc1. *Cell, 150*(6), 1196-1208. doi:https://doi.org/10.1016/j.cell.2012.07.032

Bar-Peled, L., Schweitzer, L. D., Zoncu, R., & Sabatini, D. M. (2012). Ragulator is a gef for the rag gtpases that signal amino acid levels to mtorc1. *Cell, 150*(6), 1196-1208. doi:10.1016/j.cell.2012.07.032

Belyy, V., Schlager, M. A., Foster, H., Reimer, A. E., Carter, A. P., & Yildiz, A. (2016). The mammalian dynein-dynactin complex is a strong opponent to kinesin in a tug-of-war competition. *Nat Cell Biol, 18*(9), 1018-1024. doi:10.1038/ncb3393

Binotti, B., Pavlos, N. J., Riedel, D., Wenzel, D., Vorbrüggen, G., Schalk, A. M., . . . Jahn, R. (2015). The gtpase rab26 links synaptic vesicles to the autophagy pathway. *Elife, 4*, e05597-e05597. doi:10.7554/eLife.05597

Bjørkøy, G., Lamark, T., Brech, A., Outzen, H., Perander, M., Overvatn, A., . . . Johansen, T. (2005). P62/sqstm1 forms protein aggregates degraded by autophagy and has a protective effect on huntingtin-induced cell death. *The Journal of Cell Biology, 171*(4), 603-614. doi:10.1083/jcb.200507002

Blommaart, E. F., Krause, U., Schellens, J. P., Vreeling-Sindelarova, H., & Meijer, A. J. (1997). The phosphatidylinositol 3-kinase inhibitors wortmannin and ly294002 inhibit autophagy in isolated rat hepatocytes. *Eur J Biochem, 243*(1-2), 240-246. doi:10.1111/j.1432-1033.1997.0240a.x

Bolhy, S., Bouhlel, I., Dultz, E., Nayak, T., Zuccolo, M., Gatti, X., . . . Doye, V. (2011). A nup133-dependent npc-anchored network tethers centrosomes to the nuclear envelope in prophase. *J Cell Biol, 192*(5), 855-871.

Bordi, M., Berg, M. J., Mohan, P. S., Peterhoff, C. M., Alldred, M. J., Che, S., . . . Nixon, R. A. (2016). Autophagy flux in ca1 neurons of alzheimer hippocampus: Increased induction overburdens failing lysosomes to propel neuritic dystrophy. *Autophagy, 12*(12), 2467-2483. doi:10.1080/15548627.2016.1239003

Brewer, G. J., & Cotman, C. W. (1989). Survival and growth of hippocampal neurons in defined medium at low density: Advantages of a sandwich culture technique or low oxygen. *Brain Res, 494*(1), 65-74. doi:10.1016/0006-8993(89)90144-3

Brewer, G. J., Torricelli, J. R., Evege, E. K., & Price, P. J. (1993). Optimized survival of hippocampal neurons in b27-supplemented neurobasal, a new serum-free medium combination. *J Neurosci Res, 35*(5), 567-576. doi:10.1002/jnr.490350513

Bright, N. A., Davis, L. J., & Luzio, J. P. (2016). Endolysosomes are the principal intracellular sites of acid hydrolase activity. *Curr Biol, 26*(17), 2233-2245. doi:10.1016/j.cub.2016.06.046

Bucci, C., De Gregorio, L., & Bruni, C. B. (2001). Expression analysis and chromosomal assignment of pra1 and rilp genes. *Biochem Biophys Res Commun, 286*(4), 815-819. doi:10.1006/bbrc.2001.5466

Budovskaya, Y. V., Stephan, J. S., Deminoff, S. J., & Herman, P. K. (2005). An evolutionary proteomics approach identifies substrates of the camp-dependent protein kinase. *Proceedings of the National Academy of Sciences of the United States of America, 102*(39), 13933-13938. doi:10.1073/pnas.0501046102

Caccamo, A., Ferreira, E., Branca, C., & Oddo, S. (2017a). P62 improves ad-like pathology by increasing autophagy. *Molecular psychiatry, 22*(6), 865-873. doi:10.1038/mp.2016.139

Caccamo, A., Ferreira, E., Branca, C., & Oddo, S. (2017b). P62 improves ad-like pathology by increasing autophagy. *Mol Psychiatry, 22*(6), 865-873. doi:10.1038/mp.2016.139

Caccamo, A., Majumder, S., Richardson, A., Strong, R., & Oddo, S. (2010). Molecular interplay between mammalian target of rapamycin (mtor), amyloid-beta, and tau: Effects on cognitive impairments. *J Biol Chem, 285*(17), 13107-13120. doi:10.1074/jbc.M110.100420

Cantalupo, G., Alifano, P., Roberti, V., Bruni, C. B., & Bucci, C. (2001). Rab-interacting lysosomal protein (rilp): The rab7 effector required for transport to lysosomes. *Embo j, 20*(4), 683-693. doi:10.1093/emboj/20.4.683

Carlsson, S. R., & Simonsen, A. (2015). Membrane dynamics in autophagosome biogenesis. *J Cell Sci, 128*(2), 193-205. doi:10.1242/jcs.141036

Cecconi, F., & Levine, B. (2008). The role of autophagy in mammalian development: Cell makeover rather than cell death. *Dev Cell, 15*(3), 344-357. doi:10.1016/j.devcel.2008.08.012

Celestino, R., Henen, M. A., Gama, J. B., Carvalho, C., McCabe, M., Barbosa, D. J., . . . Vogeli, B. (2019). A transient helix in the disordered region of dynein light intermediate chain links the motor to structurally diverse adaptors for cargo transport. *PLoS Biol, 17*(1), e3000100. doi:10.1371/journal.pbio.3000100

Celestino, R., Henen, M. A., Gama, J. B., Carvalho, C., McCabe, M., Barbosa, D. J., . . . Vögeli, B. (2019). A transient helix in the disordered region of dynein light intermediate chain links the motor to structurally diverse adaptors for cargo transport. *PLOS Biology, 17*(1), e3000100. doi:10.1371/journal.pbio.3000100

Chan, E. Y. W., Longatti, A., McKnight, N. C., & Tooze, S. A. (2009). Kinase-inactivated ulk proteins inhibit autophagy via their conserved c-terminal domains using an atg13-independent mechanism. *Molecular and Cellular Biology, 29*(1), 157-171. doi:10.1128/mcb.01082-08

Chang, J., Lee, S., & Blackstone, C. (2014). Spastic paraplegia proteins spastizin and spatacsin mediate autophagic lysosome reformation. *The Journal of Clinical Investigation, 124*(12), 5249-5262. doi:10.1172/JCI77598

Chantranupong, L., Scaria, S. M., Saxton, R. A., Gygi, M. P., Shen, K., Wyant, G. A., . . . Sabatini, D. M. (2016). The castor proteins are arginine sensors for the mtorc1 pathway. *Cell, 165*(1), 153-164. doi:10.1016/j.cell.2016.02.035

Chantranupong, L., Wolfson, R. L., Orozco, J. M., Saxton, R. A., Scaria, S. M., Bar-Peled, L., . . . Sabatini, D. M. (2014). The sestrins interact with gator2 to negatively regulate the amino-acid-sensing pathway upstream of mtorc1. *Cell Rep, 9*(1), 1-8. doi:10.1016/j.celrep.2014.09.014

Chen, L., Hu, J., Yun, Y., & Wang, T. (2010). Rab36 regulates the spatial distribution of late endosomes and lysosomes through a similar mechanism to rab34. *Mol Membr Biol, 27*(1), 23-30. doi:10.3109/09687680903417470

Chen, Y., Stevens, B., Chang, J., Milbrandt, J., Barres, B. A., & Hell, J. W. (2008). Ns21: Redefined and modified supplement b27 for neuronal cultures. *J Neurosci Methods, 171*(2), 239-247. doi:10.1016/j.jneumeth.2008.03.013

Cheng, J., Fujita, A., Yamamoto, H., Tatematsu, T., Kakuta, S., Obara, K., . . . Fujimoto, T. (2014). Yeast and mammalian autophagosomes exhibit distinct phosphatidylinositol 3-phosphate asymmetries. *Nature Communications, 5*(1), 3207. doi:10.1038/ncomms4207

Cheng, X.-T., Zhou, B., Lin, M.-Y., Cai, Q., & Sheng, Z.-H. (2015). Axonal autophagosomes recruit dynein for retrograde transport through fusion with late endosomes. *The Journal of Cell Biology, 209*(3), 377-386. doi:10.1083/jcb.201412046

Cheng, X. T., Zhou, B., Lin, M. Y., Cai, Q., & Sheng, Z. H. (2015). Axonal autophagosomes recruit dynein for retrograde transport through fusion with late endosomes. *J Cell Biol, 209*(3), 377-386. doi:10.1083/jcb.201412046

Cheung, P. C., Trinkle-Mulcahy, L., Cohen, P., & Lucocq, J. M. (2001). Characterization of a novel phosphatidylinositol 3-phosphate-binding protein containing two fyve fingers in tandem that is targeted to the golgi. *Biochem J, 355*(Pt 1), 113-121. doi:10.1042/0264-6021:3550113

Chklovskii, D. B. (2004). Synaptic connectivity and neuronal morphology: Two sides of the same coin. *Neuron, 43*(5), 609-617. doi:10.1016/j.neuron.2004.08.012

Crino, P. B. (2013). Evolving neurobiology of tuberous sclerosis complex. *Acta Neuropathol, 125*(3), 317-332. doi:10.1007/s00401-013-1085-x

Curatolo, P., & Moavero, R. (2012). Mtor inhibitors in tuberous sclerosis complex. *Current neuropharmacology, 10*(4), 404-415. doi:10.2174/157015912804143595

D'Costa, Vanessa M., Braun, V., Landekic, M., Shi, R., Proteau, A., McDonald, L., . . . Brumell, John H. (2015). <em>salmonella</em> disrupts host endocytic trafficking by sopd2-mediated inhibition of rab7. *Cell Reports, 12*(9), 1508-1518. doi:10.1016/j.celrep.2015.07.063

Danieli, A., & Martens, S. (2018). P62-mediated phase separation at the intersection of the ubiquitin-proteasome system and autophagy. *J Cell Sci, 131*(19), jcs214304. doi:10.1242/jcs.214304

De Luca, M., Cogli, L., Progida, C., Nisi, V., Pascolutti, R., Sigismund, S., . . . Bucci, C. (2014). Rilp regulates vacuolar atpase through interaction with the v1g1 subunit. *J Cell Sci, 127*(Pt 12), 2697-2708. doi:10.1242/jcs.142604

de Mello, N. P., Orellana, A. M., Mazucanti, C. H., de Morais Lima, G., Scavone, C., & Kawamoto, E. M. (2019). Insulin and autophagy in neurodegeneration. *Frontiers in Neuroscience, 13*, 491-491. doi:10.3389/fnins.2019.00491

De Pace, R., Skirzewski, M., Damme, M., Mattera, R., Mercurio, J., Foster, A. M., . . . Bonifacino, J. S. (2018). Altered distribution of atg9a and accumulation of axonal aggregates in neurons from a mouse model of ap-4 deficiency syndrome. *PLOS Genetics, 14*(4), e1007363. doi:10.1371/journal.pgen.1007363

Dehay, B., Bové, J., Rodríguez-Muela, N., Perier, C., Recasens, A., Boya, P., & Vila, M. (2010). Pathogenic lysosomal depletion in parkinson's disease. *J Neurosci, 30*(37), 12535-12544. doi:10.1523/jneurosci.1920-10.2010

Derubeis, A. R., Young, M. F., Jia, L., Robey, P. G., & Fisher, L. W. (2000). Double fyve-containing protein 1 (dfcp1): Isolation, cloning and characterization of a novel fyve finger protein from a human bone marrow cdna library. *Gene, 255*(2), 195-203.

Dibble, C. C., & Cantley, L. C. (2015). Regulation of mtorc1 by pi3k signaling. *Trends Cell Biol, 25*(9), 545-555. doi:10.1016/j.tcb.2015.06.002

Dickson, D. W., Braak, H., Duda, J. E., Duyckaerts, C., Gasser, T., Halliday, G. M., . . . Litvan, I. (2009). Neuropathological assessment of parkinson's disease: Refining the diagnostic criteria. *Lancet Neurol, 8*(12), 1150-1157. doi:10.1016/s1474-4422(09)70238-8

Dikic, I., & Elazar, Z. (2018). Mechanism and medical implications of mammalian autophagy. *Nat Rev Mol Cell Biol, 19*(6), 349-364. doi:10.1038/s41580-018-0003-4

Ding, W. X., Li, M., Chen, X., Ni, H. M., Lin, C. W., Gao, W., . . . Yin, X. M. (2010). Autophagy reduces acute ethanol-induced hepatotoxicity and steatosis in mice. *Gastroenterology, 139*(5), 1740-1752. doi:10.1053/j.gastro.2010.07.041

Dix, C. I., Soundararajan, H. C., Dzhindzhev, N. S., Begum, F., Suter, B., Ohkura, H., . . . Bullock, S. L. (2013). Lissencephaly-1 promotes the recruitment of dynein and dynactin to transported mrnas. *J Cell Biol, 202*(3), 479-494. doi:10.1083/jcb.201211052

Dooley, H. C., Razi, M., Polson, H. E., Girardin, S. E., Wilson, M. I., & Tooze, S. A. (2014). Wipi2 links lc3 conjugation with pi3p, autophagosome formation, and pathogen clearance by recruiting atg12-5-16l1. *Mol Cell, 55*(2), 238-252. doi:10.1016/j.molcel.2014.05.021

Du, W., Su, Qian P., Chen, Y., Zhu, Y., Jiang, D., Rong, Y., . . . Yu, L. (2016). Kinesin 1 drives autolysosome tubulation. *Developmental Cell, 37*(4), 326-336. doi:https://doi.org/10.1016/j.devcel.2016.04.014

Efimov, V. P., & Morris, N. R. (2000). The lis1-related nudf protein of aspergillus nidulans interacts with the coiled-coil domain of the nude/ro11 protein. *J Cell Biol, 150*(3), 681-688.

Egan, M. J., Tan, K., & Reck-Peterson, S. L. (2012). Lis1 is an initiation factor for dynein-driven organelle transport. *J Cell Biol, 197*(7), 971-982. doi:10.1083/jcb.201112101

Fader, C. M., Sánchez, D. G., Mestre, M. B., & Colombo, M. I. (2009). Ti-vamp/vamp7 and vamp3/cellubrevin: Two v-snare proteins involved in specific steps of the autophagy/multivesicular body pathways. *Biochimica et Biophysica Acta (BBA) - Molecular Cell Research, 1793*(12), 1901-1916. doi:https://doi.org/10.1016/j.bbamcr.2009.09.011

Feng, Y., He, D., Yao, Z., & Klionsky, D. J. (2014). The machinery of macroautophagy. *Cell research, 24*(1), 24-41. doi:10.1038/cr.2013.168

Feng, Y., Olson, E. C., Stukenberg, P. T., Flanagan, L. A., Kirschner, M. W., & Walsh, C. A. (2000). Lis1 regulates cns lamination by interacting with mnude, a central component of the centrosome. *Neuron, 28*(3), 665-679.

Filimonenko , M., Stuffers , S., Raiborg , C., Yamamoto , A., Malerød , L., Fisher , E. M. C., . . . Simonsen , A. (2007). Functional multivesicular bodies are required for autophagic clearance of protein aggregates associated with neurodegenerative disease. *The Journal of Cell Biology, 179*(3), 485-500. doi:10.1083/jcb.200702115

Forgac, M. (2007). Vacuolar atpases: Rotary proton pumps in physiology and pathophysiology. *Nat Rev Mol Cell Biol, 8*(11), 917-929. doi:10.1038/nrm2272

Fu, M. M., Nirschl, J. J., & Holzbaur, E. L. F. (2014). Lc3 binding to the scaffolding protein jip1 regulates processive dynein-driven transport of autophagosomes. *Dev Cell, 29*(5), 577-590. doi:10.1016/j.devcel.2014.04.015

Gama, J. B., Pereira, C., Simoes, P. A., Celestino, R., Reis, R. M., Barbosa, D. J., . . . Gassmann, R. (2017). Molecular mechanism of dynein recruitment to kinetochores by the rod-zw10-zwilch complex and spindly. *J Cell Biol, 216*(4), 943-960. doi:10.1083/jcb.201610108

Gammoh, N., Florey, O., Overholtzer, M., & Jiang, X. (2013). Interaction between fip200 and atg16l1 distinguishes ulk1 complex-dependent and -independent autophagy. *Nat Struct Mol Biol, 20*(2), 144-149. doi:10.1038/nsmb.2475

Gan, X., Wang, J., Su, B., & Wu, D. (2011). Evidence for direct activation of mtorc2 kinase activity by phosphatidylinositol 3,4,5-trisphosphate. *J Biol Chem, 286*(13), 10998-11002. doi:10.1074/jbc.M110.195016

Gao, J., Langemeyer, L., Kummel, D., Reggiori, F., & Ungermann, C. (2018). Molecular mechanism to target the endosomal mon1-ccz1 gef complex to the pre-autophagosomal structure. *Elife, 7*. doi:10.7554/eLife.31145

Ge, L., Melville, D., Zhang, M., & Schekman, R. (2013). The er-golgi intermediate compartment is a key membrane source for the lc3 lipidation step of autophagosome biogenesis. *Elife, 2*, e00947. doi:10.7554/eLife.00947

Gowrishankar, S., Yuan, P., Wu, Y., Schrag, M., Paradise, S., Grutzendler, J., . . . Ferguson, S. M. (2015). Massive accumulation of luminal protease-deficient axonal lysosomes at alzheimer's disease amyloid plaques. *Proc Natl Acad Sci U S A, 112*(28), E3699-3708. doi:10.1073/pnas.1510329112

Gowrishankar, S., Yuan, P., Wu, Y., Schrag, M., Paradise, S., Grutzendler, J., . . . Ferguson, S. M. (2015). Massive accumulation of luminal protease-deficient axonal lysosomes at alzheimer's disease amyloid plaques. *Proceedings of the National Academy of Sciences, 112*(28), E3699-E3708. doi:10.1073/pnas.1510329112

Guignot, J., Caron, E., Beuzon, C., Bucci, C., Kagan, J., Roy, C., & Holden, D. W. (2004). Microtubule motors control membrane dynamics of salmonella-containing vacuoles. *J Cell Sci, 117*(Pt 7), 1033-1045. doi:10.1242/jcs.00949

Guillén, C., & Benito, M. (2018). Mtorc1 overactivation as a key aging factor in the progression to type 2 diabetes mellitus. *Frontiers in endocrinology, 9*, 621-621. doi:10.3389/fendo.2018.00621

Gutierrez, P. A., Ackermann, B. E., Vershinin, M., & McKenney, R. J. (2017). Differential effects of the dynein-regulatory factor lissencephaly-1 on processive dynein-dynactin motility. *J Biol Chem, 292*(29), 12245-12255. doi:10.1074/jbc.M117.790048

Hamasaki, M., Furuta, N., Matsuda, A., Nezu, A., Yamamoto, A., Fujita, N., . . . Yoshimori, T. (2013). Autophagosomes form at er–mitochondria contact sites. *Nature, 495*(7441), 389-393. doi:10.1038/nature11910

Hara, T., Nakamura, K., Matsui, M., Yamamoto, A., Nakahara, Y., Suzuki-Migishima, R., . . . Mizushima, N. (2006). Suppression of basal autophagy in neural cells causes neurodegenerative disease in mice. *Nature, 441*(7095), 885-889. doi:10.1038/nature04724

Harrison, R. E., Brumell, J. H., Khandani, A., Bucci, C., Scott, C. C., Jiang, X., . . . Grinstein, S. (2004). Salmonella impairs rilp recruitment to rab7 during maturation of invasion vacuoles. *Mol Biol Cell, 15*(7), 3146-3154. doi:10.1091/mbc.e04-02-0092

Harrison, R. E., Bucci, C., Vieira, O. V., Schroer, T. A., & Grinstein, S. (2003). Phagosomes fuse with late endosomes and/or lysosomes by extension of membrane protrusions along microtubules: Role of rab7 and rilp. *Mol Cell Biol, 23*(18), 6494-6506.

Hayashi-Nishino, M., Fujita, N., Noda, T., Yamaguchi, A., Yoshimori, T., & Yamamoto, A. (2009). A subdomain of the endoplasmic reticulum forms a cradle for autophagosome formation. *Nat Cell Biol, 11*(12), 1433-1437. doi:10.1038/ncb1991

Heck, N., & Benavides-Piccione, R. (2015). Editorial: Dendritic spines: From shape to function. *Frontiers in neuroanatomy, 9*, 101-101. doi:10.3389/fnana.2015.00101

Hernandez, D., Torres, C. A., Setlik, W., Cebrián, C., Mosharov, E. V., Tang, G., . . . Sulzer, D. (2012). Regulation of presynaptic neurotransmission by macroautophagy. *Neuron, 74*(2), 277-284. doi:10.1016/j.neuron.2012.02.020

Heuser, J. (1989). Changes in lysosome shape and distribution correlated with changes in cytoplasmic ph. *J Cell Biol, 108*(3), 855-864. doi:10.1083/jcb.108.3.855

Hollenbeck, P. J. (1993). Products of endocytosis and autophagy are retrieved from axons by regulated retrograde organelle transport. *J Cell Biol, 121*(2), 305-315. doi:10.1083/jcb.121.2.305

Hoogenraad, C. C., & Akhmanova, A. (2016). Bicaudal d family of motor adaptors: Linking dynein motility to cargo binding. *Trends Cell Biol*. doi:10.1016/j.tcb.2016.01.001

Hoogenraad, C. C., Akhmanova, A., Howell, S. A., Dortland, B. R., De Zeeuw, C. I., Willemsen, R., . . . Galjart, N. (2001). Mammalian golgi-associated bicaudal-d2 functions in the dynein-dynactin pathway by interacting with these complexes. *EMBO J, 20*(15), 4041-4054. doi:10.1093/emboj/20.15.4041

Hu, D. J., Baffet, A. D., Nayak, T., Akhmanova, A., Doye, V., & Vallee, R. B. (2013). Dynein recruitment to nuclear pores activates apical nuclear migration and mitotic entry in brain progenitor cells. *Cell, 154*(6), 1300-1313. doi:10.1016/j.cell.2013.08.024

Huang, J., Roberts, A. J., Leschziner, A. E., & Reck-Peterson, S. L. (2012). Lis1 acts as a "clutch" between the atpase and microtubule-binding domains of the dynein motor. *Cell, 150*(5), 975-986. doi:10.1016/j.cell.2012.07.022

Hübner, C. A., & Dikic, I. (2020). Er-phagy and human diseases. *Cell Death & Differentiation, 27*(3), 833-842. doi:10.1038/s41418-019-0444-0

Huynh, W., & Vale, R. D. (2017). Disease-associated mutations in human bicd2 hyperactivate motility of dynein-dynactin. *J Cell Biol, 216*(10), 3051-3060. doi:10.1083/jcb.201703201

Itoh, T., Fujita, N., Kanno, E., Yamamoto, A., Yoshimori, T., & Fukuda, M. (2008). Golgi-resident small gtpase rab33b interacts with atg16l and modulates autophagosome formation. *Mol Biol Cell, 19*(7), 2916-2925. doi:10.1091/mbc.e07-12-1231

Jahreiss, L., Menzies, F. M., & Rubinsztein, D. C. (2008). The itinerary of autophagosomes: From peripheral formation to kiss-and-run fusion with lysosomes. *Traffic, 9*(4), 574-587. doi:10.1111/j.1600-0854.2008.00701.x

Jang, M., Park, R., Kim, H., Namkoong, S., Jo, D., Huh, Y. H., . . . Park, J. (2018). Ampk contributes to autophagosome maturation and lysosomal fusion. *Scientific Reports, 8*(1), 12637. doi:10.1038/s41598-018-30977-7

Jia, R., Guardia, C. M., Pu, J., Chen, Y., & Bonifacino, J. S. (2017). Borc coordinates encounter and fusion of lysosomes with autophagosomes. *Autophagy, 13*(10), 1648-1663. doi:10.1080/15548627.2017.1343768

Johnston, J. A., Ward, C. L., & Kopito, R. R. (1998). Aggresomes: A cellular response to misfolded proteins. *J Cell Biol, 143*(7), 1883-1898. doi:10.1083/jcb.143.7.1883

Jordens, I., Fernandez-Borja, M., Marsman, M., Dusseljee, S., Janssen, L., Calafat, J., . . . Neefjes, J. (2001). The rab7 effector protein rilp controls lysosomal transport by inducing the recruitment of dynein-dynactin motors. *Curr Biol, 11*(21), 1680-1685.

Kallergi, E., Daskalaki, A.-D., Ioannou, E., Kolaxi, A., Plataki, M., Haberkant, P., . . . Nikoletopoulou, V. (2020). Long-term synaptic depression triggers local biogenesis of autophagic vesicles in dendrites and requires autophagic degradation. *bioRxiv*, 2020.2003.2012.983965. doi:10.1101/2020.03.12.983965

Kalvari, I., Tsompanis, S., Mulakkal, N. C., Osgood, R., Johansen, T., Nezis, I. P., & Promponas, V. J. (2014). Ilir: A web resource for prediction of atg8-family interacting proteins. *Autophagy, 10*(5), 913-925. doi:10.4161/auto.28260

Kaushik, S., & Cuervo, A. M. (2018). The coming of age of chaperone-mediated autophagy. *Nature Reviews Molecular Cell Biology, 19*(6), 365-381. doi:10.1038/s41580-018-0001-6

Kaushik, S., & Cuervo, A. M. (2018). The coming of age of chaperone-mediated autophagy. *Nat Rev Mol Cell Biol, 19*(6), 365-381. doi:10.1038/s41580-018-0001-6

Khaminets, A., Behl, C., & Dikic, I. (2016). Ubiquitin-dependent and independent signals in selective autophagy. *Trends Cell Biol, 26*(1), 6-16. doi:10.1016/j.tcb.2015.08.010

Kihara, A., Noda, T., Ishihara, N., & Ohsumi, Y. (2001). Two distinct vps34 phosphatidylinositol 3-kinase complexes function in autophagy and carboxypeptidase y sorting in saccharomyces cerevisiae. *J Cell Biol, 152*(3), 519-530. doi:10.1083/jcb.152.3.519

Kihara, A., Noda, T., Ishihara, N., & Ohsumi, Y. (2001). Two distinct vps34 phosphatidylinositol 3–kinase complexes function in autophagy and carboxypeptidase y sorting insaccharomyces cerevisiae. *The Journal of Cell Biology, 152*(3), 519-530. doi:10.1083/jcb.152.3.519

Kim, H. J., Hong, Y. B., Park, J.-M., Choi, Y.-R., Kim, Y. J., Yoon, B. R., . . . Choi, B.-O. (2013). Mutations in the plekhg5 gene is relevant with autosomal recessive intermediate charcot-marie-tooth disease. *Orphanet Journal of Rare Diseases, 8*(1), 104. doi:10.1186/1750-1172-8-104

Kim, J., Kundu, M., Viollet, B., & Guan, K.-L. (2011). Ampk and mtor regulate autophagy through direct phosphorylation of ulk1. *Nature Cell Biology, 13*, 132. doi:10.1038/ncb2152 <u>https://www.nature.com/articles/ncb2152 - supplementary-information</u>

Kim, J., Kundu, M., Viollet, B., & Guan, K. L. (2011). Ampk and mtor regulate autophagy through direct phosphorylation of ulk1. *Nat Cell Biol, 13*(2), 132-141. doi:10.1038/ncb2152

Klionsky, D. J. (2005). The molecular machinery of autophagy: Unanswered questions. *J Cell Sci, 118*(Pt 1), 7-18. doi:10.1242/jcs.01620

Knævelsrud, H., Søreng, K., Raiborg, C., Håberg, K., Rasmuson, F., Brech, A., . . . Simonsen, A. (2013). Membrane remodeling by the px-bar protein snx18 promotes autophagosome formation. *J Cell Biol, 202*(2), 331-349. doi:10.1083/jcb.201205129

Kononenko, N. L., Claßen, G. A., Kuijpers, M., Puchkov, D., Maritzen, T., Tempes, A., . . . Haucke, V. (2017). Retrograde transport of trkb-containing autophagosomes via the adaptor ap-2 mediates neuronal complexity and prevents neurodegeneration. *Nature Communications, 8*(1), 14819. doi:10.1038/ncomms14819

Korolchuk, V. I., Saiki, S., Lichtenberg, M., Siddiqi, F. H., Roberts, E. A., Imarisio, S., . . . Rubinsztein, D. C. (2011). Lysosomal positioning coordinates cellular nutrient responses. *Nat Cell Biol, 13*(4), 453-460. doi:10.1038/ncb2204

Kounakis, K., Chaniotakis, M., Markaki, M., & Tavernarakis, N. (2019). Emerging roles of lipophagy in health and disease. *Frontiers in Cell and Developmental Biology, 7*(185). doi:10.3389/fcell.2019.00185

Koyama-Honda, I., Itakura, E., Fujiwara, T. K., & Mizushima, N. (2013). Temporal analysis of recruitment of mammalian atg proteins to the autophagosome formation site. *Autophagy, 9*(10), 1491-1499. doi:10.4161/auto.25529

Kraft, C., Reggiori, F., & Peter, M. (2009). Selective types of autophagy in yeast. *Biochim Biophys Acta, 1793*(9), 1404-1412. doi:10.1016/j.bbamcr.2009.02.006

Kubicki, M., Park, H., Westin, C. F., Nestor, P. G., Mulkern, R. V., Maier, S. E., . . . Shenton, M. E. (2005). Dti and mtr abnormalities in schizophrenia: Analysis of white matter integrity. *Neuroimage, 26*(4), 1109-1118. doi:10.1016/j.neuroimage.2005.03.026

Kudchodkar, S. B., & Levine, B. (2009). Viruses and autophagy. *Rev Med Virol, 19*(6), 359-378. doi:10.1002/rmv.630

Kuusisto, E., Salminen, A., & Alafuzoff, I. (2002). Early accumulation of p62 in neurofibrillary tangles in alzheimer's disease: Possible role in tangle formation. *Neuropathology and Applied Neurobiology, 28*(3), 228-237. doi:10.1046/j.1365-2990.2002.00394.x

Kwon, C. H., Luikart, B. W., Powell, C. M., Zhou, J., Matheny, S. A., Zhang, W., . . . Parada, L. F. (2006). Pten regulates neuronal arborization and social interaction in mice. *Neuron, 50*(3), 377-388. doi:10.1016/j.neuron.2006.03.023

Lamark, T., Kirkin, V., Dikic, I., & Johansen, T. (2009). Nbr1 and p62 as cargo receptors for selective autophagy of ubiquitinated targets. *Cell Cycle, 8*(13), 1986-1990. doi:10.4161/cc.8.13.8892

Lamb, C. A., Yoshimori, T., & Tooze, S. A. (2013). The autophagosome: Origins unknown, biogenesis complex. *Nat Rev Mol Cell Biol, 14*(12), 759-774. doi:10.1038/nrm3696

Lee, I. G., Olenick, M. A., Boczkowska, M., Franzini-Armstrong, C., Holzbaur, E. L. F., & Dominguez, R. (2018). A conserved interaction of the dynein light intermediate chain with dynein-dynactin effectors necessary for processivity. *Nat Commun, 9*(1), 986. doi:10.1038/s41467-018-03412-8

Li, M., Hou, Y., Wang, J., Chen, X., Shao, Z. M., & Yin, X. M. (2011). Kinetics comparisons of mammalian atg4 homologues indicate selective preferences toward diverse atg8 substrates. *J Biol Chem, 286*(9), 7327-7338. doi:10.1074/jbc.M110.199059

Li, W.-w., Li, J., & Bao, J.-k. (2012). Microautophagy: Lesser-known self-eating. *Cellular and Molecular Life Sciences, 69*(7), 1125-1136. doi:10.1007/s00018-011-0865-5

Lieberman, O. J., Frier, M. D., McGuirt, A. F., Griffey, C. J., Rafikian, E., Yang, M., . . . Sulzer, D. (2020). Cell-type-specific regulation of neuronal intrinsic excitability by macroautophagy. *Elife, 9*. doi:10.7554/eLife.50843

Lin, X., Yang, T., Wang, S., Wang, Z., Yun, Y., Sun, L., . . . Wang, T. (2014). Rilp interacts with hops complex via vps41 subunit to regulate endocytic trafficking. *Sci Rep, 4*, 7282. doi:10.1038/srep07282

Lindmo, K., & Stenmark, H. (2006). Regulation of membrane traffic by phosphoinositide 3-kinases. *J Cell Sci, 119*(Pt 4), 605-614. doi:10.1242/jcs.02855

Liu, J., & Li, L. (2019). Targeting autophagy for the treatment of alzheimer's disease: Challenges and opportunities. *Frontiers in Molecular Neuroscience, 12*(203). doi:10.3389/fnmol.2019.00203

Longatti, A., Lamb, C. A., Razi, M., Yoshimura, S., Barr, F. A., & Tooze, S. A. (2012). Tbc1d14 regulates autophagosome formation via rab11- and ulk1-positive recycling endosomes. *J Cell Biol, 197*(5), 659-675. doi:10.1083/jcb.201111079

Lüningschrör, P., Binotti, B., Dombert, B., Heimann, P., Perez-Lara, A., Slotta, C., . . . Sendtner, M. (2017). Plekhg5-regulated autophagy of synaptic vesicles reveals a pathogenic mechanism in motoneuron disease. *Nature Communications, 8*(1), 678. doi:10.1038/s41467-017-00689-z

Luo, Z., Saha, A. K., Xiang, X., & Ruderman, N. B. (2005). Ampk, the metabolic syndrome and cancer. *Trends Pharmacol Sci, 26*(2), 69-76. doi:10.1016/j.tips.2004.12.011

Ma, S., Attarwala, I. Y., & Xie, X.-Q. (2019). Sqstm1/p62: A potential target for neurodegenerative disease. *ACS chemical neuroscience, 10*(5), 2094-2114. doi:10.1021/acschemneuro.8b00516

MacIntosh, G. C., & Bassham, D. C. (2011). The connection between ribophagy, autophagy and ribosomal rna decay. *Autophagy, 7*(6), 662-663. doi:10.4161/auto.7.6.15447

Maday, S., & Holzbaur, E. L. F. (2016). Compartment-specific regulation of autophagy in primary neurons. *J Neurosci, 36*(22), 5933-5945. doi:10.1523/jneurosci.4401-15.2016

Martinez-Vicente, M., Talloczy, Z., Wong, E., Tang, G., Koga, H., Kaushik, S., . . . Cuervo, A. M. (2010). Cargo recognition failure is responsible for inefficient autophagy in huntington's disease. *Nat Neurosci, 13*(5), 567-576. doi:10.1038/nn.2528

Matanis, T., Akhmanova, A., Wulf, P., Del Nery, E., Weide, T., Stepanova, T., . . . Hoogenraad, C. C. (2002). Bicaudal-d regulates copi-independent golgi-er transport by recruiting the dynein-dynactin motor complex. *Nat Cell Biol, 4*(12), 986-992.

Mattera, R., Park, S. Y., De Pace, R., Guardia, C. M., & Bonifacino, J. S. (2017). Ap-4 mediates export of atg9a from the <em>trans</em>-golgi network to promote autophagosome formation. *Proceedings of the National Academy of Sciences, 114*(50), E10697-E10706. doi:10.1073/pnas.1717327114

Mauvezin, C., & Neufeld, T. P. (2015). Bafilomycin a1 disrupts autophagic flux by inhibiting both v-atpase-dependent acidification and ca-p60a/serca-dependent autophagosome-lysosome fusion. *Autophagy, 11*(8), 1437-1438. doi:10.1080/15548627.2015.1066957

Mavrakis, M., Lippincott-Schwartz, J., Stratakis, C. A., & Bossis, I. (2006). Depletion of type ia regulatory subunit (riα) of protein kinase a (pka) in mammalian cells and tissues activates mtor and causes autophagic deficiency. *Human Molecular Genetics, 15*(19), 2962-2971. doi:10.1093/hmg/ddl239

McKenney, R. J., Huynh, W., Tanenbaum, M. E., Bhabha, G., & Vale, R. D. (2014). Activation of cytoplasmic dynein motility by dynactin-cargo adapter complexes. *Science, 345*(6194), 337-341. doi:10.1126/science.1254198 science.1254198 [pii]

McKenney, R. J., Vershinin, M., Kunwar, A., Vallee, R. B., & Gross, S. P. (2010). Lis1 and nude induce a persistent dynein force-producing state. *Cell, 141*(2), 304-314.

Millecamps, S., & Julien, J.-P. (2013). Axonal transport deficits and neurodegenerative diseases. *Nature Reviews Neuroscience, 14*(3), 161-176. doi:10.1038/nrn3380

Mizushima, N. (2018). A brief history of autophagy from cell biology to physiology and disease. *Nat Cell Biol, 20*(5), 521-527. doi:10.1038/s41556-018-0092-5

Mizushima, N., Yoshimori, T., & Ohsumi, Y. (2011). The role of atg proteins in autophagosome formation. *Annu Rev Cell Dev Biol, 27*, 107-132. doi:10.1146/annurev-cellbio-092910-154005

Mochizuki, Y., Ohashi, R., Kawamura, T., Iwanari, H., Kodama, T., Naito, M., & Hamakubo, T. (2013). Phosphatidylinositol 3-phosphatase myotubularin-related protein 6 (mtmr6) is regulated by small gtpase rab1b in the early secretory and autophagic pathways. *J Biol Chem, 288*(2), 1009-1021. doi:10.1074/jbc.M112.395087

Moors, T., Paciotti, S., Chiasserini, D., Calabresi, P., Parnetti, L., Beccari, T., & van de Berg, W. D. (2016). Lysosomal dysfunction and α-synuclein aggregation in parkinson's disease: Diagnostic links. *Mov Disord, 31*(6), 791-801. doi:10.1002/mds.26562

Moreau, K., Ravikumar, B., Renna, M., Puri, C., & Rubinsztein, David C. (2011). Autophagosome precursor maturation requires homotypic fusion. *Cell, 146*(2), 303-317. doi:10.1016/j.cell.2011.06.023

Moriya, S., Komatsu, S., Yamasaki, K., Kawai, Y., Kokuba, H., Hirota, A., . . . Miyazawa, K. (2015). Targeting the integrated networks of aggresome formation, proteasome, and autophagy potentiates er stress-mediated cell death in multiple myeloma cells. *Int J Oncol, 46*(2), 474-486. doi:10.3892/ijo.2014.2773

Munson, M. J., Allen, G. F., Toth, R., Campbell, D. G., Lucocq, J. M., & Ganley, I. G. (2015). Mtor activates the vps34–uvrag complex to regulate autolysosomal tubulation and cell survival. *The EMBO Journal, 34*(17), 2272-2290. doi:10.15252/embj.201590992

Murphy, K. E., Gysbers, A. M., Abbott, S. K., Spiro, A. S., Furuta, A., Cooper, A., . . . Halliday, G. M. (2015). Lysosomal-associated membrane protein 2 isoforms are differentially affected in early parkinson's disease. *Movement Disorders, 30*(12), 1639-1647. doi:10.1002/mds.26141

Narendra, D., Kane, L. A., Hauser, D. N., Fearnley, I. M., & Youle, R. J. (2010). P62/sqstm1 is required for parkin-induced mitochondrial clustering but not mitophagy; vdac1 is dispensable for both. *Autophagy, 6*(8), 1090-1106.

Narendra, D., Tanaka, A., Suen, D. F., & Youle, R. J. (2008). Parkin is recruited selectively to impaired mitochondria and promotes their autophagy. *J Cell Biol, 183*(5), 795-803. doi:10.1083/jcb.200809125

Nie, D., Di Nardo, A., Han, J. M., Baharanyi, H., Kramvis, I., Huynh, T., . . . Sahin, M. (2010). Tsc2-rheb signaling regulates epha-mediated axon guidance. *Nat Neurosci, 13*(2), 163-172. doi:10.1038/nn.2477

Niethammer, M., Smith, D. S., Ayala, R., Peng, J., Ko, J., Lee, M. S., . . . Tsai, L. H. (2000). Nudel is a novel cdk5 substrate that associates with lis1 and cytoplasmic dynein. *Neuron, 28*(3), 697-711.

Nikoletopoulou, V., Sidiropoulou, K., Kallergi, E., Dalezios, Y., & Tavernarakis, N. (2017). Modulation of autophagy by bdnf underlies synaptic plasticity. *Cell Metab, 26*(1), 230-242.e235. doi:10.1016/j.cmet.2017.06.005

Nikoletopoulou, V., Sidiropoulou, K., Kallergi, E., Dalezios, Y., & Tavernarakis, N. (2017). Modulation of autophagy by bdnf underlies synaptic plasticity. *Cell Metabolism, 26*(1), 230-242.e235. doi:https://doi.org/10.1016/j.cmet.2017.06.005

Nikoletopoulou, V., & Tavernarakis, N. (2018). Regulation and roles of autophagy at synapses. *Trends Cell Biol, 28*(8), 646-661. doi:10.1016/j.tcb.2018.03.006

Nilsson, P., Loganathan, K., Sekiguchi, M., Matsuba, Y., Hui, K., Tsubuki, S., . . . Saido, Takaomi C. (2013). Aβ secretion and plaque formation depend on autophagy. *Cell Reports, 5*(1), 61-69. doi:https://doi.org/10.1016/j.celrep.2013.08.042

Nilsson, P., & Saido, T. C. (2014). Dual roles for autophagy: Degradation and secretion of alzheimer's disease aβ peptide. *BioEssays : news and reviews in molecular, cellular and developmental biology, 36*(6), 570-578. doi:10.1002/bies.201400002

Nishimura, T., Kaizuka, T., Cadwell, K., Sahani, M. H., Saitoh, T., Akira, S., . . . Mizushima, N. (2013). Fip200 regulates targeting of atg16l1 to the isolation membrane. *EMBO reports, 14*(3), 284-291. doi:10.1038/embor.2013.6

Nixon, R. A., Wegiel, J., Kumar, A., Yu, W. H., Peterhoff, C., Cataldo, A., & Cuervo, A. M. (2005). Extensive involvement of autophagy in alzheimer disease: An immuno-electron microscopy study. *Journal of Neuropathology & Experimental Neurology, 64*(2), 113-122. doi:10.1093/jnen/64.2.113

Nixon, R. A., Wegiel, J., Kumar, A., Yu, W. H., Peterhoff, C., Cataldo, A., & Cuervo, A. M. (2005). Extensive involvement of autophagy in alzheimer disease: An immuno-electron microscopy study. *J Neuropathol Exp Neurol, 64*(2), 113-122. doi:10.1093/jnen/64.2.113

Ntsapi, C., Lumkwana, D., Swart, C., du Toit, A., & Loos, B. (2018). New insights into autophagy dysfunction related to amyloid beta toxicity and neuropathology in alzheimer's disease. *Int Rev Cell Mol Biol, 336*, 321-361. doi:10.1016/bs.ircmb.2017.07.002

Ohsumi, Y. (2014). Historical landmarks of autophagy research. *Cell research, 24*(1), 9-23. doi:10.1038/cr.2013.169

Orsi, A., Razi, M., Dooley, H. C., Robinson, D., Weston, A. E., Collinson, L. M., & Tooze, S. A. (2012). Dynamic and transient interactions of atg9 with autophagosomes, but not membrane integration, are required for autophagy. *Molecular Biology of the Cell, 23*(10), 1860-1873. doi:10.1091/mbc.e11-09-0746

Pandey, J. P., & Smith, D. S. (2011). A cdk5-dependent switch regulates lis1/ndel1/dynein-driven organelle transport in adult axons. *Journal of Neuroscience, 31*(47), 17207-17219.

Pankiv, S., Alemu, E. A., Brech, A., Bruun, J. A., Lamark, T., Overvatn, A., . . . Johansen, T. (2010). Fyco1 is a rab7 effector that binds to lc3 and pi3p to mediate microtubule plus end-directed vesicle transport. *J Cell Biol, 188*(2), 253-269. doi:10.1083/jcb.200907015

Papassotiropoulos, A., Bagli, M., Kurz, A., Kornhuber, J., Förstl, H., Maier, W., . . . Heun, R. (2000). A genetic variation of cathepsin d is a major risk factor for alzheimer's disease. *Annals of Neurology, 47*(3), 399-403. doi:10.1002/1531-8249(200003)47:3<399::AID-ANA22>3.0.CO;2-5

Paschal, B. M., & Vallee, R. B. (1987). Retrograde transport by the microtubule associated protein map 1c. *Nature, 330*, 181-183.

Progida, C., Malerod, L., Stuffers, S., Brech, A., Bucci, C., & Stenmark, H. (2007). Rilp is required for the proper morphology and function of late endosomes. *J Cell Sci, 120*(Pt 21), 3729-3737. doi:10.1242/jcs.017301

Progida, C., Spinosa, M. R., De Luca, A., & Bucci, C. (2006). Rilp interacts with the vps22 component of the escrt-ii complex. *Biochem Biophys Res Commun, 347*(4), 1074-1079. doi:10.1016/j.bbrc.2006.07.007

Qi, L., Zhang, X. D., Wu, J. C., Lin, F., Wang, J., DiFiglia, M., & Qin, Z. H. (2012). The role of chaperone-mediated autophagy in huntingtin degradation. *PLoS One, 7*(10), e46834. doi:10.1371/journal.pone.0046834

Raiborg, C., Wenzel, E. M., Pedersen, N. M., Olsvik, H., Schink, K. O., Schultz, S. W., . . . Stenmark, H. (2015). Repeated er-endosome contacts promote endosome translocation and neurite outgrowth. *Nature, 520*(7546), 234-238. doi:10.1038/nature14359

Reddy, B. J., Mattson, M., Wynne, C. L., Vadpey, O., Durra, A., Chapman, D., . . . Gross, S. P. (2016). Load-induced enhancement of dynein force production by lis1-nude in vivo and in vitro. *Nat Commun, 7*, 12259. doi:10.1038/ncomms12259

Reef, S., Zalckvar, E., Shifman, O., Bialik, S., Sabanay, H., Oren, M., & Kimchi, A. (2006). A short mitochondrial form of p19arf induces autophagy and caspase-independent cell death. *Mol Cell, 22*(4), 463-475. doi:10.1016/j.molcel.2006.04.014

Ribeiro, M., López de Figueroa, P., Blanco, F. J., Mendes, A. F., & Caramés, B. (2016). Insulin decreases autophagy and leads to cartilage degradation. *Osteoarthritis and Cartilage, 24*(4), 731-739. doi:https://doi.org/10.1016/j.joca.2015.10.017

Richter, B., Sliter, D. A., Herhaus, L., Stolz, A., Wang, C., Beli, P., . . . Dikic, I. (2016). Phosphorylation of optn by tbk1 enhances its binding to ub chains and promotes selective autophagy of damaged mitochondria. *Proc Natl Acad Sci U S A, 113*(15), 4039-4044. doi:10.1073/pnas.1523926113

Roczniak-Ferguson, A., Petit, C. S., Froehlich, F., Qian, S., Ky, J., Angarola, B., . . . Ferguson, S. M. (2012). The transcription factor tfeb links mtorc1 signaling to transcriptional control of lysosome homeostasis. *Sci Signal, 5*(228), ra42. doi:10.1126/scisignal.2002790

Roczniak-Ferguson, A., Petit, C. S., Froehlich, F., Qian, S., Ky, J., Angarola, B., . . . Ferguson, S. M. (2012). The transcription factor tfeb links mtorc1 signaling to transcriptional control of lysosome homeostasis. *Sci Signal, 5*(228), ra42. doi:10.1126/scisignal.2002790

Rogov, V., Dotsch, V., Johansen, T., & Kirkin, V. (2014). Interactions between autophagy receptors and ubiquitin-like proteins form the molecular basis for selective autophagy. *Mol Cell, 53*(2), 167-178. doi:10.1016/j.molcel.2013.12.014

Rong, Y., McPhee, C. K., Deng, S., Huang, L., Chen, L., Liu, M., . . . Lenardo, M. J. (2011). Spinster is required for autophagic lysosome reformation and mtor reactivation following starvation. *Proceedings of the National Academy of Sciences, 108*(19), 7826-7831. doi:10.1073/pnas.1013800108

Ross, C. A., Margolis, R. L., Reading, S. A., Pletnikov, M., & Coyle, J. T. (2006). Neurobiology of schizophrenia. *Neuron, 52*(1), 139-153. doi:10.1016/j.neuron.2006.09.015

Rowland, A. M., Richmond, J. E., Olsen, J. G., Hall, D. H., & Bamber, B. A. (2006). Presynaptic terminals independently regulate synaptic clustering and autophagy of gabaa receptors in caenorhabditis elegans. *J Neurosci, 26*(6), 1711-1720. doi:10.1523/jneurosci.2279-05.2006

Russell, R. C., Tian, Y., Yuan, H., Park, H. W., Chang, Y. Y., Kim, J., . . . Guan, K. L. (2013). Ulk1 induces autophagy by phosphorylating beclin-1 and activating vps34 lipid kinase. *Nat Cell Biol, 15*(7), 741-750. doi:10.1038/ncb2757

Saillour, Y., Carion, N., Quelin, C., Leger, P. L., Boddaert, N., Elie, C., . . . Bahi-Buisson, N. (2009). Lis1-related isolated lissencephaly: Spectrum of mutations and relationships with malformation severity. *Arch Neurol, 66*(8), 1007-1015. doi:10.1001/archneurol.2009.149

Sancak, Y., Peterson, T. R., Shaul, Y. D., Lindquist, R. A., Thoreen, C. C., Bar-Peled, L., & Sabatini, D. M. (2008). The rag gtpases bind raptor and mediate amino acid signaling to mtorc1. *Science, 320*(5882), 1496-1501. doi:10.1126/science.1157535

Sarkar, S., Ravikumar, B., & Rubinsztein, D. C. (2009). Autophagic clearance of aggregate-prone proteins associated with neurodegeneration. *Methods Enzymol, 453*, 83-110. doi:10.1016/s0076-6879(08)04005-6

Sasaki, S., Shionoya, A., Ishida, M., Gambello, M. J., Yingling, J., Wynshaw-Boris, A., & Hirotsune, S. (2000). A lis1/nudel/cytoplasmic dynein heavy chain complex in the developing and adult nervous system. *Neuron, 28*(3), 681-696.

Sato, A. (2016). Mtor, a potential target to treat autism spectrum disorder. *CNS Neurol Disord Drug Targets, 15*(5), 533-543. doi:10.2174/1871527315666160413120638

Saxton, R. A., Chantranupong, L., Knockenhauer, K. E., Schwartz, T. U., & Sabatini, D. M. (2016). Mechanism of arginine sensing by castor1 upstream of mtorc1. *Nature, 536*(7615), 229-233. doi:10.1038/nature19079

Saxton, R. A., & Sabatini, D. M. (2017). Mtor signaling in growth, metabolism, and disease. *Cell, 168*(6), 960-976. doi:10.1016/j.cell.2017.02.004

Scherer, J., Yi, J., & Vallee, R. B. (2014). Pka-dependent dynein switching from lysosomes to adenovirus: A novel form of host-virus competition. *J Cell Biol, 205*(2), 163-177. doi:10.1083/jcb.201307116

Schlager, M. A., Hoang, H. T., Urnavicius, L., Bullock, S. L., & Carter, A. P. (2014). In vitro reconstitution of a highly processive recombinant human dynein complex. *EMBO J, 33*(17), 1855-1868. doi:10.15252/embj.201488792 embj.201488792 [pii]

Schulze, R. J., Weller, S. G., Schroeder, B., Krueger, E. W., Chi, S., Casey, C. A., & McNiven, M. A. (2013). Lipid droplet breakdown requires dynamin 2 for vesiculation of autolysosomal tubules in hepatocytes. *The Journal of Cell Biology, 203*(2), 315-326. doi:10.1083/jcb.201306140

Seto, S., Matsumoto, S., Tsujimura, K., & Koide, Y. (2010). Differential recruitment of cd63 and rab7-interacting-lysosomal-protein to phagosomes containing mycobacterium tuberculosis in macrophages. *Microbiol Immunol, 54*(3), 170-174. doi:10.1111/j.1348-0421.2010.00199.x

Settembre, C., Zoncu, R., Medina, D. L., Vetrini, F., Erdin, S., Erdin, S., . . . Ballabio, A. (2012). A lysosome-to-nucleus signalling mechanism senses and regulates the lysosome via mtor and tfeb. *EMBO J, 31*(5), 1095-1108. doi:10.1038/emboj.2012.32

Shehata, M., Matsumura, H., Okubo-Suzuki, R., Ohkawa, N., & Inokuchi, K. (2012). Neuronal stimulation induces autophagy in hippocampal neurons that is involved in ampa receptor degradation after chemical long-term depression. *J Neurosci, 32*(30), 10413-10422. doi:10.1523/jneurosci.4533-11.2012

Shu, T., Ayala, R., Nguyen, M. D., Xie, Z., Gleeson, J. G., & Tsai, L. H. (2004). Ndel1 operates in a common pathway with lis1 and cytoplasmic dynein to regulate cortical neuronal positioning. *Neuron, 44*(2), 263-277.

Singh, R., Kaushik, S., Wang, Y., Xiang, Y., Novak, I., Komatsu, M., . . . Czaja, M. J. (2009). Autophagy regulates lipid metabolism. *Nature, 458*(7242), 1131-1135. doi:10.1038/nature07976

Slobodkin, M. R., & Elazar, Z. (2013). The atg8 family: Multifunctional ubiquitin-like key regulators of autophagy. *Essays Biochem, 55*, 51-64. doi:10.1042/bse0550051

Soreng, K., Munson, M. J., Lamb, C. A., Bjorndal, G. T., Pankiv, S., Carlsson, S. R., . . . Simonsen, A. (2018). Snx18 regulates atg9a trafficking from recycling endosomes by recruiting dynamin-2. *EMBO Rep, 19*(4). doi:10.15252/embr.201744837

Soukup, S. F., Kuenen, S., Vanhauwaert, R., Manetsberger, J., Hernández-Díaz, S., Swerts, J., . . . Verstreken, P. (2016). A lrrk2-dependent endophilina phosphoswitch is critical for macroautophagy at presynaptic terminals. *Neuron, 92*(4), 829-844. doi:10.1016/j.neuron.2016.09.037

Spencer, K. M., Nestor, P. G., Niznikiewicz, M. A., Salisbury, D. F., Shenton, M. E., & McCarley, R. W. (2003). Abnormal neural synchrony in schizophrenia. *J Neurosci, 23*(19), 7407-7411. doi:10.1523/jneurosci.23-19-07407.2003

Splinter, D., Razafsky, D. S., Schlager, M. A., Serra-Marques, A., Grigoriev, I., Demmers, J., . . . Akhmanova, A. (2012). Bicd2, dynactin, and lis1 cooperate in regulating dynein recruitment to cellular structures. *Mol Biol Cell, 23*(21), 4226-4241. doi:10.1091/mbc.E12-03-0210 mbc.E12-03-0210 [pii]

Splinter, D., Tanenbaum, M. E., Lindqvist, A., Jaarsma, D., Flotho, A., Yu, K. L., . . . Akhmanova, A. (2010). Bicaudal d2, dynein, and kinesin-1 associate with nuclear pore complexes and regulate centrosome and nuclear positioning during mitotic entry. *PLoS Biol, 8*(4), e1000350. doi:10.1371/journal.pbio.1000350

Sridhar, S., Patel, B., Aphkhazava, D., Macian, F., Santambrogio, L., Shields, D., & Cuervo, A. M. (2013). The lipid kinase pi4kiiiβ preserves lysosomal identity. *EMBO J, 32*(3), 324-339. doi:10.1038/emboj.2012.341

Starling, G. P., Yip, Y. Y., Sanger, A., Morton, P. E., Eden, E. R., & Dodding, M. P. (2016). Folliculin directs the formation of a rab34–rilp complex to control the nutrient-dependent dynamic distribution of lysosomes. *EMBO reports, 17*(6), 823-841. doi:10.15252/embr.201541382

Starr, T., Ng, T. W., Wehrly, T. D., Knodler, L. A., & Celli, J. (2008). Brucella intracellular replication requires trafficking through the late endosomal/lysosomal compartment. *Traffic, 9*(5), 678-694. doi:10.1111/j.1600-0854.2008.00718.x

Stehman, S. A., Chen, Y., McKenney, R. J., & Vallee, R. B. (2007). Nude and nudel are required for mitotic progression and are involved in dynein recruitment to kinetochores. *J Cell Biol, 178*(4), 583-594. doi:jcb.200610112 [pii]
10.1083/jcb.200610112

Stephan, J. S., Yeh, Y.-Y., Ramachandran, V., Deminoff, S. J., & Herman, P. K. (2009a). The tor and pka signaling pathways independently target the atg1/atg13 protein kinase complex to control autophagy. *Proceedings of the National Academy of Sciences, 106*(40), 17049-17054. doi:10.1073/pnas.0903316106

Stephan, J. S., Yeh, Y.-Y., Ramachandran, V., Deminoff, S. J., & Herman, P. K. (2009b). The tor and pka signaling pathways independently target the atg1/atg13 protein kinase complex to control autophagy. *Proceedings of the National Academy of Sciences of the United States of America, 106*(40), 17049-17054. doi:10.1073/pnas.0903316106

Stolz, A., Ernst, A., & Dikic, I. (2014). Cargo recognition and trafficking in selective autophagy. *Nat Cell Biol, 16*(6), 495-501. doi:10.1038/ncb2979

Sun, J., Deghmane, A. E., Soualhine, H., Hong, T., Bucci, C., Solodkin, A., & Hmama, Z. (2007). Mycobacterium bovis bcg disrupts the interaction of rab7 with rilp contributing to inhibition of phagosome maturation. *J Leukoc Biol, 82*(6), 1437-1445. doi:10.1189/jlb.10.1189

Sun-Wada, G. H., Wada, Y., & Futai, M. (2004). Diverse and essential roles of mammalian vacuolar-type proton pump atpase: Toward the physiological understanding of inside acidic compartments. *Biochim Biophys Acta, 1658*(1-2), 106-114. doi:10.1016/j.bbabio.2004.04.013

Takahashi, Y., He, H., Tang, Z., Hattori, T., Liu, Y., Young, M. M., . . . Wang, H.-G. (2018). An autophagy assay reveals the escrt-iii component chmp2a as a regulator of phagophore closure. *Nature Communications, 9*(1), 2855. doi:10.1038/s41467-018-05254-w

Takahashi, Y., Meyerkord, C. L., Hori, T., Runkle, K., Fox, T. E., Kester, M., . . . Wang, H. G. (2011). Bif-1 regulates atg9 trafficking by mediating the fission of golgi membranes during autophagy. *Autophagy, 7*(1), 61-73. doi:10.4161/auto.7.1.14015

Takei, N., & Nawa, H. (2014). Mtor signaling and its roles in normal and abnormal brain development. *Front Mol Neurosci, 7*, 28. doi:10.3389/fnmol.2014.00028

Tan, S. C., Scherer, J., & Vallee, R. B. (2011). Recruitment of dynein to late endosomes and lysosomes through light intermediate chains. *Mol Biol Cell, 22*(4), 467-477. doi:10.1091/mbc.E10-02-0129

Tang, G., Gudsnuk, K., Kuo, S.-H., Cotrina, M. L., Rosoklija, G., Sosunov, A., . . . Sulzer, D. (2014). Loss of mtor-dependent macroautophagy causes autistic-like synaptic pruning deficits. *Neuron, 83*(5), 1131-1143. doi:10.1016/j.neuron.2014.07.040

Tang, G., Gudsnuk, K., Kuo, S. H., Cotrina, M. L., Rosoklija, G., Sosunov, A., . . . Sulzer, D. (2014). Loss of mtor-dependent macroautophagy causes autistic-like synaptic pruning deficits. *Neuron, 83*(5), 1131-1143. doi:10.1016/j.neuron.2014.07.040

Thoreen, C. C., & Sabatini, D. M. (2009). Rapamycin inhibits mtorc1, but not completely. *Autophagy, 5*(5), 725-726. doi:8504 [pii]

Toropova, K., Zou, S., Roberts, A. J., Redwine, W. B., Goodman, B. S., Reck-Peterson, S. L., & Leschziner, A. E. (2014). Lis1 regulates dynein by sterically blocking its mechanochemical cycle. *Elife, 3*. doi:10.7554/eLife.03372

Tramutola, A., Triplett, J. C., Di Domenico, F., Niedowicz, D. M., Murphy, M. P., Coccia, R., . . . Butterfield, D. A. (2015). Alteration of mtor signaling occurs early in the progression of alzheimer disease (ad): Analysis of brain from subjects with pre-clinical ad, amnestic mild cognitive impairment and late-stage ad. *Journal of Neurochemistry, 133*(5), 739-749. doi:10.1111/jnc.13037

Tsai, J. W., Bremner, K. H., & Vallee, R. B. (2007). Dual subcellular roles for lis1 and dynein in radial neuronal migration in live brain tissue. *Nat Neurosci, 10*(8), 970-979.

Tsai, J. W., Chen, Y., Kriegstein, A. R., & Vallee, R. B. (2005). Lis1 rna interference blocks neural stem cell division, morphogenesis, and motility at multiple stages. *J Cell Biol, 170*(6), 935-945.

Tsai, J. W., Lian, W. N., Kemal, S., Kriegstein, A. R., & Vallee, R. B. (2010). Kinesin 3 and cytoplasmic dynein mediate interkinetic nuclear migration in neural stem cells. *Nat Neurosci, 13*(12), 1463-1471.

Tsun, Z. Y., Bar-Peled, L., Chantranupong, L., Zoncu, R., Wang, T., Kim, C., . . . Sabatini, D. M. (2013). The folliculin tumor suppressor is a gap for the ragc/d gtpases that signal amino acid levels to mtorc1. *Mol Cell, 52*(4), 495-505. doi:10.1016/j.molcel.2013.09.016

Uemura, T., Yamamoto, M., Kametaka, A., Sou, Y.-s., Yabashi, A., Yamada, A., . . . Waguri, S. (2014). A cluster of thin tubular structures mediates transformation of the endoplasmic reticulum to autophagic isolation membrane. *Molecular and Cellular Biology, 34*(9), 1695-1706. doi:10.1128/mcb.01327-13

Um, J.-H., & Yun, J. (2017). Emerging role of mitophagy in human diseases and physiology. *BMB reports, 50*(6), 299-307. doi:10.5483/bmbrep.2017.50.6.056

Urnavicius, L., Lau, C. K., Elshenawy, M. M., Morales-Rios, E., Motz, C., Yildiz, A., & Carter, A. P. (2018). Cryo-em shows how dynactin recruits two dyneins for faster movement. *Nature, 554*(7691), 202-206. doi:10.1038/nature25462

Urnavicius, L., Zhang, K., Diamant, A. G., Motz, C., Schlager, M. A., Yu, M., . . . Carter, A. P. (2015). The structure of the dynactin complex and its interaction with dynein. *Science, 347*(6229), 1441-1446. doi:10.1126/science.aaa4080

Vallee, R. B., Wall, J. S., Paschal, B. M., & Shpetner, H. S. (1988). Microtubule associated protein 1c from brain is a two-headed cytosolic dynein. *Nature, 332*, 561-563.

Vallee, R. B., Wall, J. S., Paschal, B. M., & Shpetner, H. S. (1988). Microtubule-associated protein 1c from brain is a two-headed cytosolic dynein. *Nature, 332*(6164), 561-563. doi:10.1038/332561a0

van der Kant, R., Fish, A., Janssen, L., Janssen, H., Krom, S., Ho, N., . . . Neefjes, J. (2013). Late endosomal transport and tethering are coupled processes controlled by rilp and the cholesterol sensor orp1l. *Journal of Cell Science, 126*(15), 3462-3474. doi:10.1242/jcs.129270

Vander Haar, E., Lee, S. I., Bandhakavi, S., Griffin, T. J., & Kim, D. H. (2007). Insulin signalling to mtor mediated by the akt/pkb substrate pras40. *Nat Cell Biol, 9*(3), 316-323. doi:10.1038/ncb1547

Varga, R.-E., Khundadze, M., Damme, M., Nietzsche, S., Hoffmann, B., Stauber, T., . . . Hübner, C. A. (2015). In vivo evidence for lysosome depletion and impaired autophagic clearance in hereditary spastic paraplegia type spg11. *PLOS Genetics, 11*(8), e1005454. doi:10.1371/journal.pgen.1005454

Wang, S., Tsun, Z.-Y., Wolfson, R. L., Shen, K., Wyant, G. A., Plovanich, M. E., . . . Sabatini, D. M. (2015). Lysosomal amino acid transporter slc38a9 signals arginine sufficiency to mtorc1. *Science, 347*(6218), 188. doi:10.1126/science.1257132

Wang, T., & Hong, W. (2005). Assay and functional properties of rab34 interaction with rilp in lysosome morphogenesis. *Methods Enzymol, 403*, 675-687. doi:10.1016/s0076-6879(05)03058-2

Wang, T., & Hong, W. (2006). Rilp interacts with vps22 and vps36 of escrt-ii and regulates their membrane recruitment. *Biochem Biophys Res Commun, 350*(2), 413-423. doi:10.1016/j.bbrc.2006.09.064

Wang, T., Wong, K. K., & Hong, W. (2004). A unique region of rilp distinguishes it from its related proteins in its regulation of lysosomal morphology and interaction with rab7 and rab34. *Mol Biol Cell, 15*(2), 815-826. doi:10.1091/mbc.e03-06-0413

Wang, Y., Martinez-Vicente, M., Krüger, U., Kaushik, S., Wong, E., Mandelkow, E.-M., . . . Mandelkow, E. (2009). Tau fragmentation, aggregation and clearance: The dual role of lysosomal processing. *Human Molecular Genetics, 18*(21), 4153-4170. doi:10.1093/hmg/ddp367

Wijdeven, R. H., Janssen, H., Nahidiazar, L., Janssen, L., Jalink, K., Berlin, I., & Neefjes, J. (2016). Cholesterol and orp1l-mediated er contact sites control autophagosome transport and fusion with the endocytic pathway. *Nat Commun, 7*, 11808. doi:10.1038/ncomms11808

Wolfson, R. L., Chantranupong, L., Saxton, R. A., Shen, K., Scaria, S. M., Cantor, J. R., & Sabatini, D. M. (2016). Sestrin2 is a leucine sensor for the mtorc1 pathway. *Science, 351*(6268), 43-48. doi:10.1126/science.aab2674

Wolfson, R. L., Chantranupong, L., Wyant, G. A., Gu, X., Orozco, J. M., Shen, K., . . . Sabatini, D. M. (2017). Kicstor recruits gator1 to the lysosome and is necessary for nutrients to regulate mtorc1. *Nature, 543*(7645), 438-442. doi:10.1038/nature21423

Wozniak, A. L., Long, A., Jones-Jamtgaard, K. N., & Weinman, S. A. (2016). Hepatitis c virus promotes virion secretion through cleavage of the rab7 adaptor protein rilp. *Proc Natl Acad Sci U S A, 113*(44), 12484-12489. doi:10.1073/pnas.1607277113

Wu, M., Wang, T., Loh, E., Hong, W., & Song, H. (2005). Structural basis for recruitment of rilp by small gtpase rab7. *EMBO J, 24*(8), 1491-1501. doi:10.1038/sj.emboj.7600643

Wurzer, B., Zaffagnini, G., Fracchiolla, D., Turco, E., Abert, C., Romanov, J., & Martens, S. (2015). Oligomerization of p62 allows for selection of ubiquitinated cargo and isolation membrane during selective autophagy. *Elife, 4*, e08941. doi:10.7554/eLife.08941

Wyant, G. A., Abu-Remaileh, M., Wolfson, R. L., Chen, W. W., Freinkman, E., Danai, L. V., . . . Sabatini, D. M. (2017). Mtorc1 activator slc38a9 is required to efflux essential amino acids from lysosomes and use protein as a nutrient. *Cell, 171*(3), 642-654.e612. doi:https://doi.org/10.1016/j.cell.2017.09.046

Yamamoto, A., Cremona, M. L., & Rothman, J. E. (2006). Autophagy-mediated clearance of huntingtin aggregates triggered by the insulin-signaling pathway. *Journal of Cell Biology, 172*(5), 719-731. doi:10.1083/jcb.200510065

Yamamoto, S., Kuramoto, K., Wang, N., Situ, X., Priyadarshini, M., Zhang, W., . . . He, C. (2018). Autophagy differentially regulates insulin production and insulin sensitivity. *Cell Reports, 23*(11), 3286-3299. doi:10.1016/j.celrep.2018.05.032

Yap, C. C., Digilio, L., McMahon, L. P., Garcia, A. D. R., & Winckler, B. (2018a). Degradation of dendritic cargos requires rab7-dependent transport to somatic lysosomes. *J Cell Biol*. doi:10.1083/jcb.201711039

Yap, C. C., Digilio, L., McMahon, L. P., Garcia, A. D. R., & Winckler, B. (2018b). Degradation of dendritic cargos requires rab7-dependent transport to somatic lysosomes. *J Cell Biol, 217*(9), 3141-3159. doi:10.1083/jcb.201711039

Yasuda, S., Morishita, S., Fujita, A., Nanao, T., Wada, N., Waguri, S., . . . Nakamura, T. (2016). Mon1–ccz1 activates rab7 only on late endosomes and dissociates from the lysosome in mammalian cells. *Journal of Cell Science, 129*(2), 329. doi:10.1242/jcs.178095

Yi, J. Y., Ori-McKenney, K. M., McKenney, R. J., Vershinin, M., Gross, S. P., & Vallee, R. B. (2011). High-resolution imaging reveals indirect coordination of opposite motors and a role for lis1 in high-load axonal transport. *J Cell Biol, 195*(2), 193-201. doi:10.1083/jcb.201104076

Yim, W. W.-Y., & Mizushima, N. (2020). Lysosome biology in autophagy. *Cell Discovery, 6*(1), 6. doi:10.1038/s41421-020-0141-7

Ylä-Anttila, P., Vihinen, H., Jokitalo, E., & Eskelinen, E.-L. (2009). 3d tomography reveals connections between the phagophore and endoplasmic reticulum. *Autophagy, 5*(8), 1180-1185. doi:10.4161/auto.5.8.10274

Yorimitsu, T., Zaman, S., Broach, J. R., & Klionsky, D. J. (2007). Protein kinase a and sch9 cooperatively regulate induction of autophagy in saccharomyces cerevisiae. *Molecular Biology of the Cell, 18*(10), 4180-4189. doi:10.1091/mbc.e07-05-0485

Young, A. R., Chan, E. Y., Hu, X. W., Kochl, R., Crawshaw, S. G., High, S., . . . Tooze, S. A. (2006). Starvation and ulk1-dependent cycling of mammalian atg9 between the tgn and endosomes. *J Cell Sci, 119*(Pt 18), 3888-3900. doi:10.1242/jcs.03172

Yu, L., McPhee, C. K., Zheng, L., Mardones, G. A., Rong, Y., Peng, J., . . . Lenardo, M. J. (2010). Termination of autophagy and reformation of lysosomes regulated by mtor. *Nature, 465*(7300), 942-946. doi:10.1038/nature09076

Yu, W. H., Cuervo, A. M., Kumar, A., Peterhoff, C. M., Schmidt, S. D., Lee, J. H., . . . Nixon, R. A. (2005). Macroautophagy--a novel beta-amyloid peptide-generating pathway activated in alzheimer's disease. *J Cell Biol, 171*(1), 87-98. doi:10.1083/jcb.200505082

Zaffagnini, G., Savova, A., Danieli, A., Romanov, J., Tremel, S., Ebner, M., . . . Martens, S. (2018a). P62 filaments capture and present ubiquitinated cargos for autophagy. *Embo j, 37*(5). doi:10.15252/embj.201798308

Zaffagnini, G., Savova, A., Danieli, A., Romanov, J., Tremel, S., Ebner, M., . . . Martens, S. (2018b). Phasing out the bad-how sqstm1/p62 sequesters ubiquitinated proteins for degradation by autophagy. *Autophagy.* doi:10.1080/15548627.2018.1462079

Zhang, K., Foster, H. E., Rondelet, A., Lacey, S. E., Bahi-Buisson, N., Bird, A. W., & Carter, A. P. (2017). Cryo-em reveals how human cytoplasmic dynein is auto-inhibited and activated. *Cell, 169*(7), 1303-1314.e1318. doi:10.1016/j.cell.2017.05.025

Zhao, X., Nedvetsky, P., Stanchi, F., Vion, A.-C., Popp, O., Zühlke, K., . . . Gerhardt, H. (2019). Endothelial pka activity regulates angiogenesis by limiting autophagy through phosphorylation of atg16l1. *Elife, 8*, e46380. doi:10.7554/eLife.46380

Zhou, F., Wu, Z., Zhao, M., Murtazina, R., Cai, J., Zhang, A., . . . Segev, N. (2019). Rab5-dependent autophagosome closure by escrt. *The Journal of Cell Biology, 218*(6), 1908-1927. doi:10.1083/jcb.201811173

Zhu, J. H., Guo, F., Shelburne, J., Watkins, S., & Chu, C. T. (2003). Localization of phosphorylated erk/map kinases to mitochondria and autophagosomes in lewy body diseases. *Brain Pathol, 13*(4), 473-481. doi:10.1111/j.1750-3639.2003.tb00478.x